Lecture Notes in Chemistry 55

L. T. Fan D. Neogi M. Yashima

Elementary Introduction to Spatial and Temporal Fractals

Springer-Verlag
Berlin Heidelberg New York
London Paris Tokyo
Hong Kong Barcelona
Budapest

Authors

L. T. Fan
Department of Chemical Engineering
Kansas State University
Manhattan, Kansas 66506, USA

D. Neogi
Air Products and Chemicals, Inc.
7201 Hamilton Boulevard
Allentown, Pennsylvania 18195, USA

M. Yashima
Department of Statistics
Kansas State University
Manhattan, Kansas 66506, USA

ISBN-13: 978-3-540-54212-4 e-ISBN-13: 978-3-642-45690-9
DOI: 10.1007/978-3-642-45690-9

2151/3140-543210 - Printed on acid-free paper

PREFACE

This monograph presents expositionally and in an elementary manner various definitions, concepts, notions, and methodologies of both spatial and temporal fractals. Numerous simple examples and graphical illustrations are included to facilitate the presentation.

Fractal geometry dealing with spatial fractals provides both descriptions and mathematical models for a variety of the seemingly complex forms and patterns either existing in nature or artificially generated on computers. Fractal geometry plays an important role in modeling and animating natural phenomena and engineering processes, and in music and art. Unlike the topological or Euclidean dimension, the dimension of a fractal object or fractal dimension need not be an integer. B.B. Mandelbrot, who originally proposed the concept of fractals and is often referred to as the "Father" of fractal geometry, initially suggested that fractal dimension is strictly greater than the topological dimension but not greater than the Euclidean dimension; however, he later relaxed this definition to include topological dimension as the lower limit for the fractal dimension. Most natural products are irregular in shape: some of them, such as coast lines, trees, and rivers, are known to possess self-similar irregularities which can be expressed by resorting to the fractal dimensions.

When a random or stochastic process has a finite variance, the random variable characterizing the process gives rise to the celebrated Gaussian distribution. On the contrary, if a random process has an infinite variance and the distribution of its random variable is stable but non-Gaussian, its temporal structure may be fractal. The fractal time is defined as a self-similar set of time intervals between events such that the mean time between events is infinity. In order to be fractal in time, the events in a sequence must be clustered; in other words, events will occur in quick succession for a period of time, thus forming a cluster, and then there will be a long delay before another cluster of events takes place.

Chaos is a term frequently used to describe the complicated behavior of a nonlinear dynamical system, specifically, a system whose time history has a sensitive dependence on its initial condition. The concept of fractals is also an effective means for describing and quantifying certain aspects of dynamical systems exhibiting irregular or chaotic behavior. In the later part of this monograph, therefore, a brief description of the concepts of chaotic dynamics and their close connection with spatial and temporal fractals is presented.

This monograph is intended for readers uninitiated in the subject of fractals; no attempt is made to treat with mathematical rigor any of the contents. Nevertheless, the presentation is sufficiently detailed so that the readers who master any concept or methodology in the monograph with the aid of appropriate references cited can apply it to the areas of their interest with confidence.

Acknowledgement

The following individuals were participants in our research involving fractals and chaos: Dr. C.S. Chee, Professor R. Nassar, Mr. A.A. Boateng, Dr. B. Sur, and Professor W.P. Walawender; naturally the preparation of this monograph was benefitted by their participation. The authors are appreciative of Peggy Hanes' efforts in typing the manuscript. The first author (LTF) also wishes to acknowledge the aid of his wife, Eva, in preparing the manuscript. The second and third authors (DN and MY) are grateful for the emotional support provided by their parents during their research efforts.

February 1991

L.T. Fan
D. Neogi
M. Yashima

TABLE OF CONTENTS

1. INTRODUCTION

"Fractals" and "chaos" are two of the most popularized scientific or mathematical concepts and terminologies these days. They have become the most uttered words of the scientific community and even of the popular press. Fractal geometry liberates the analysis of the shapes of natural objects from the tyranny of straight lines, flat planes, and regular solids. Mandelbrot [1] has stated, "Clouds are not spheres, mountains are not cones, coastlines are not circles, and bark is not smooth, nor does lightning travel in a straight line." One way to envision fractals is to think of them as objects, each with a fractional dimension. In fact, fractals are almost the rule rather than exception in this universe. The exceptions are perfect lines, planes, and cubes.

Mandelbrot [1] has coined the word "fractal" to describe self-similar objects or to describe a family of shapes mimicking the constrained randomness found everywhere in nature. A self-similar object has the property that a part of it, if magnified, resembles the whole. The part is divisible into look-alike entities; this geometric cloning continues forever, at least, in the abstract sense of pure mathematics. Gleick [2] has characterized self-similarity as follows:

> "Self-similarity is symmetry across scale. It implies recursion, pattern inside of pattern. Mandelbrot's price charts and river charts displayed self-similarity, because not only did they produce detail at finer and finer scale, they also produced a detail with certain constant measurements. Monstrous shapes like Koch curve display self-similarity because they look exactly the same even under high magnification. The self-similarity is built into the technique of constructing the curve -- the same transformation is an easily recognizable quality. Its images are everywhere in the culture: in the infinitely deep reflection a person standing between two mirrors, or in the cartoon notion of a fish eating a small fish eating a small fish .... "

The range of natural phenomena encompassed by self-similarity is astonishing. Besides mountains, clouds, and trees, there are numerous others, including galactic clusters, turbulent flows, surface of the earth, behavior of mechanical and electrical systems, and chemical reactions. Fractal materials also include glass, colloidal aggregates, electrodeposited materials, coal and ceramic. Even inherent randomness associated with any process or phenomenon can give rise to fractals, e.g., stock market variations, fluctuating signals, or for that matter, any time series.

Moreover, the notion of time fractal has been proposed; it is associated with inter-arrival time (the interval time between events) of the renewal processes [3]. The inter-arrival time has self-similar structure and an infinite mean value. Perhaps the easiest way to explain what is meant by a fractal in time is to first explain what it is not. If events of one sort or another occur in a sequence of regular, equally spaced intervals, the sequence is the exact opposite of a fractal in time. To be a fractal in time, the sequence must have its events clustered -- happening rapidly one after the other -- with long stretches of nothingness between the clusters. More technically speaking, fractal time is a stochastic process in which the mean time between events is infinity. In practice, natural fractals are fractal in a statistical or stochastic sense, a particular shape or inter-arrival time giving no clue to the length or time scale at which it has been determined; this particular shape or inter-arrival time does not appear exactly the same at different length scales.

2. FUNDAMENTAL CONCEPTS AND DEFINITIONS

The fractal dimension is discussed by Mandelbrot [1] in his book by contrasting it to the topological or Euclidean dimension. Thus, prior to defining the fractal dimension, the topological and Euclidean dimensions will be briefly discussed [4].

Topology considers all pots with two handles to be of the same form because, if they are infinitely flexible and compressible, they can be molded into any other pot of a different shape without tearing any new opening or closing up any old one. Consequently, the topological dimensions of coastlines and circles are identical and equal to 1 because they consist of only lines without any area (Figure 2-1). A dimension theory in the Euclidean sense was developed in the nineteenth century. If a manifold of geometrical elements is defined to be a set of geometrical elements that can be labeled with a real, continuous coordinate system, its dimension, d_E, is the number of coordinates required to determine a general element of the set (Figure 2-1).

Mandelbrot [5] offers the following mathematical definition of a fractal;

> "A fractal is by definition a set for which the Hausdorff-Besicovitch dimension strictly exceeds the topological dimension."

However, Mandelbrot [6] did mention that if and when a good reason arises then the fractal set should include the subsets for which the Hausdorff dimension, d_H, is equal to the topological dimension, d_T. As such, in this monograph we will include both the Euclidean and the topological dimensions as the limiting cases for the fractal dimension, i.e.

$$d_T \leq d_F \leq d_E$$

Pfeifer and Obert [4] have stated that the difference, $d_F - d_T$ is a measure of the disorder of the system. If $d_T = d_F$, it is ordered or weakly disordered; if $d_T < d_F$, it is strongly disordered.

The above definition requires a brief introduction to Hausdorff-Besicovitch dimensions. A fractal set may be considered to be a set of points embedded in space. For example, the set of points that make up a line in an ordinary Euclidean space has the topological dimension, d_T, of 1, and also the Hausdorff-Besicovitch dimension, d_H, of 1. The fractal dimension of this line, d_F, is also 1 and thus equal to its topological dimension, d_T. Similarly, the set of points that form a surface in a space with the Euclidean dimension, d_E, of 2, has the topological dimension, d_T, of 2, and the Hausdorff-Besicovitch dimension, d_H, of 2. Note that for this "ordinary" surface,

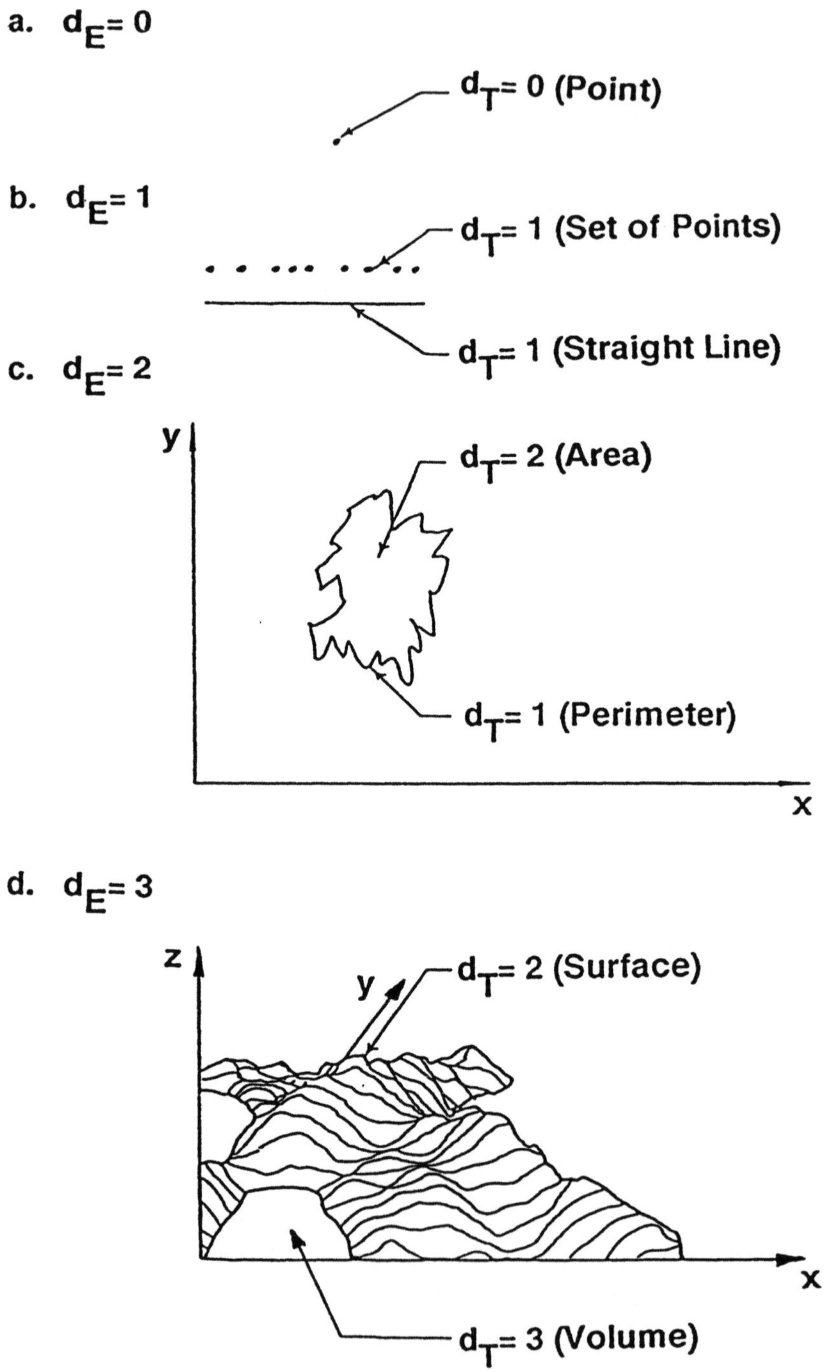

Figure 2-1. Topological and Euclidean dimensions of various forms.

$d_T = d_H = d_F = 2$. Similarly, a ball or sphere has $d_T = d_H = d_F = 3$. These shapes whose $d_T = d_H = d_F$ are commonly referred to as the "standard" or regular shapes.

The concept of a distance between points in space is central to the definition of the Hausdorff-Besicovitch dimension. The question is how we measure the "size" of a set, S, comprising points in space. This is elaborated below by borrowing heavily from Feder [7].

A simple way to measure the length of a curve or the area of a surface is by dividing the space occupied by it into either small line segments of length λ or small squares of side λ, as illustrated in Figures 2-2 and 2-3, respectively, and then, counting the number, $N(\lambda)$, of line segments or squares to cover the space.

Suppose that, for an ordinary curve, we have

$$N(\lambda) = \frac{L_0}{\lambda} \tag{2-1}$$

where L_0 is its "true" length (Figure 2-2a). The length measure, L, of the curve is given by (Figure 2-2a)

$$L = N(\lambda) \cdot \lambda = \frac{L_0}{\lambda} \cdot \lambda \longrightarrow L_0 \lambda^0 = L_0 \tag{2-2}$$

As expected the length measure, L, is equal to the true length of the curve; it is also independent of λ. Next, let us choose to associate an area with the set of points defining a curve by considering the number of squares needed to cover the curve (Figure 2-2b). The number of squares required is $N(\lambda)$, and each square has an area of λ^2. The associated area, therefore, is given by

$$A = N(\lambda) \cdot \lambda^2 = \frac{L_0}{\lambda} \cdot \lambda^2 \longrightarrow L_0 \lambda^1 \underset{\lambda \to 0}{=} 0 \tag{2-3}$$

indicating that the area measure, A, of the line approaches 0 as λ becomes increasingly small. Similarly, we may associate a volume with the line by considering the number of cubes needed to cover the line (Figure 2-2c). The associated volume is

$$V = N(\lambda) \cdot \lambda^3 = \frac{L_0}{\lambda} \cdot \lambda^3 \longrightarrow L_0 \lambda^2 \underset{\lambda \to 0}{=} 0 \tag{2-4}$$

indicating that the volume measure, V, of the line also asymptotically approaches 0 as $\lambda \longrightarrow 0$. Hence, the only meaningful measure of the set of points defined by a curve in a 2-dimensional Euclidean space is the length measure, L.

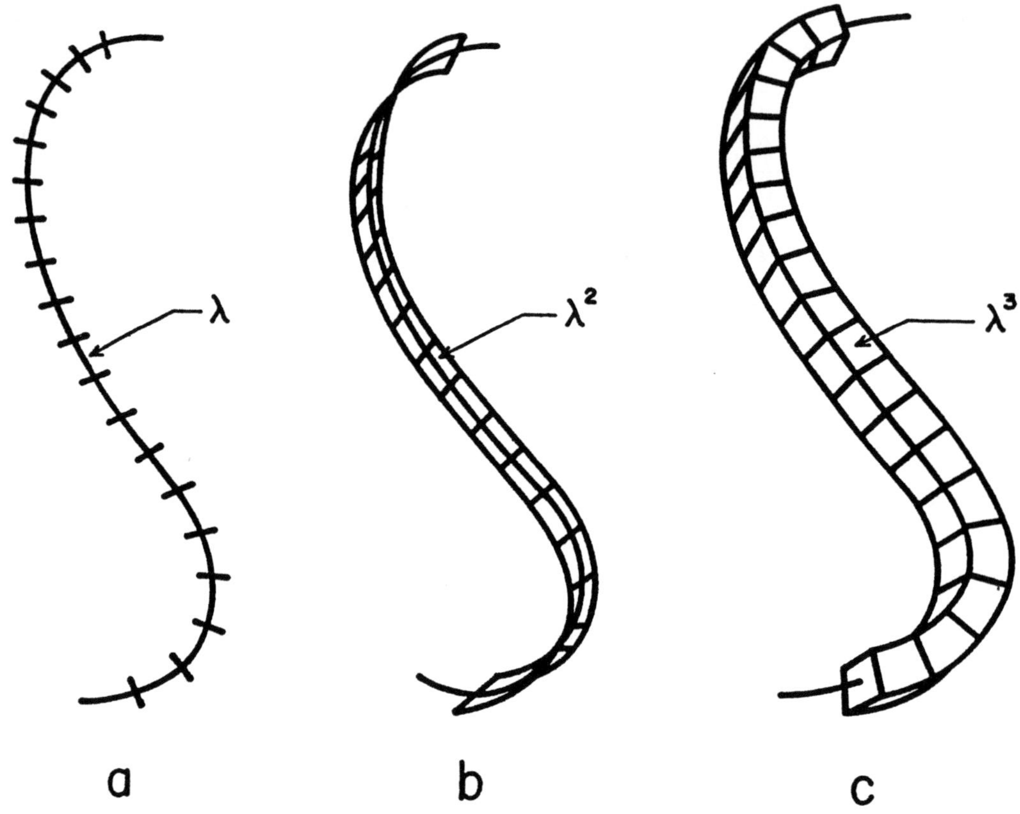

Figure 2-2. Measuring the "size" of a curve (see, e.g., [7]):

a. using line segments
b. using squares
c. using cubes.

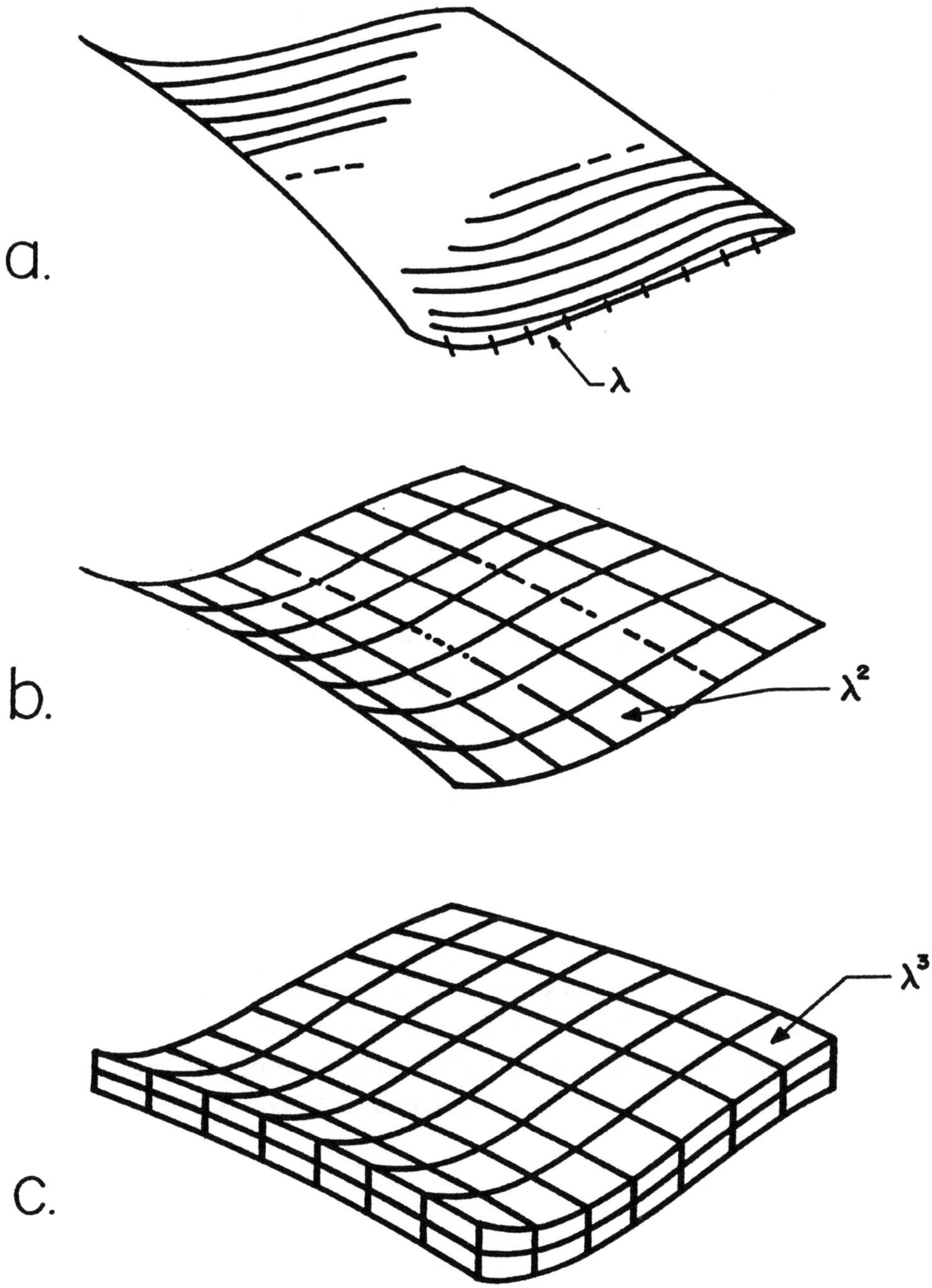

Figure 2-3. Measuring the "size" of a surface (see, e.g., [7]):

a. using line segments
b. using squares
c. using cubes.

Let us now consider a set of points that define a surface as shown in Figure 2-3. Suppose that the number of squares needed to tile an ordinary surface is

$$N(\lambda) = \frac{A_0}{\lambda^2} \tag{2-5}$$

where A_0 is the "true" area of the surface. The length measure, L, of the surface will be

$$L = N(\lambda)\cdot\lambda = \frac{A_0}{\lambda^2}\cdot\lambda \longrightarrow A_0\lambda^{-1} \underset{\lambda\to 0}{=} \infty \tag{2-6}$$

indicating that it diverges as $\lambda \longrightarrow 0$ (Figure 2-3a). This result is reasonable since it is impossible to cover a surface with a finite number of line segments. In the limit of vanishing λ, the area measure, A, of the area of the surface, as expected, will be (Figure 2-3b).

$$A = N(\lambda)\cdot\lambda^2 = \frac{A_0}{\lambda^2}\cdot\lambda^2 \longrightarrow A_0\lambda^0 = A_0 \tag{2-7}$$

In other words, the measure A is equal to the true area, A_0. We may also associate the volume measure, V, with the surface by forming the sum of the volumes of the cubes needed to cover the surface, thereby giving (Figure 2-3c)

$$V = N(\lambda)\cdot\lambda^3 = \frac{A_0}{\lambda^2}\cdot\lambda^3 \longrightarrow A_0\lambda^1 \underset{\lambda\to 0}{=} 0 \tag{2-8}$$

As expected, this volume measure, V, vanishes as $\lambda \longrightarrow 0$. Hence, the only meaningful measure of the set of points defined by a surface in a 3-dimensional Euclidean space is the area measure, A.

We may easily define a set of points that form a curve which twists so badly that its length is infinite. With the illustrations given in the preceding two paragraphs we are in a position to discuss such strange sets of points by generalizing the measures of sizes so far discussed. To give a measure of the size to a set of points, S, in space, we have considered a test function

$$h(\lambda) = \gamma(d)\lambda^d$$

representing a line, square, disk, ball or cube, whichever is used to cover the set to form the measure

$$M_d = \Sigma\, h(\lambda)$$

The geometrical factor, $\gamma(d)$, is 1 for ordinary lines, squares and cubes. Similarly, this factor is $\pi/4$ for a circular disk, and $\pi/6$ for a sphere. Note that the ordinary curve and surface illustrated in Figures 2-2 and 2-3 have $h(\lambda)$ equal to λ^1 and λ^2, respectively; the measure, M_d, is L, A or V. In general, we find that, as $\lambda \longrightarrow 0$, the measure, M_d, is infinite, finite or zero depending on the choice of d, the dimension of the measure. The Hausdorff-Besicovitch dimension of the set S, d_H, is the critical dimension for which its measure, M_d, changes from zero to infinity:

$$M_d = \Sigma\, \gamma(d)\lambda^d$$

$$= \gamma(d)N(\lambda)\lambda^d \xrightarrow[\lambda\to 0]{} \begin{Bmatrix} \infty, & d < d_H \\ 0, & d > d_H \end{Bmatrix} \qquad (2\text{-}9)$$

The example of Figure 2-2, which is an ordinary curve, is a trivial or singular case. For this example, the measure, M_1 or L, corresponding $d=1$, is finite and the measure, M_2 or A, corresponding to $d=2$ is 0 as $\lambda \to 0$; thus d_H of the curve is 1. For the example of Figure 2-3, which is an ordinary surface, the measure, M_1 or L, corresponding to $d=1$, is infinite, and the measure, M_3 or V, corresponding to $d=3$, is zero, thus, d_H which is greater than 1 and less than 3, is 2 as has been demonstrated. While d_H is a positive integer for an ordinary shape it is generally a noninteger.

Note that d_H is a local property in the sense that it measures properties of sets of points in the limit of a vanishing size λ of the test function covering the set. The dimension, d_H, is often used to estimate the fractal dimension of a set. It follows from Eq. 2-9 that asymptotically in the limit of small λ,

$$N(\lambda) \propto \frac{1}{\lambda^{d_H}} \qquad (2\text{-}10)$$

As pointed out by Barnsley [8], strictly speaking,

$$d_H \leq d_F$$

Nevertheless, instead of going into the mathematical details concerning the cases when $d_H = d_F$ and $d_H < d_F$, we refer the readers to Barnsley's monograph [8]. Also, the Hausdorff dimension, d_H, is sometimes hard to compute and its definition is not usually used as the basis of experimental procedures for the determination of fractal dimensions of physical sets.

Let us now turn our attention to a more commonly adopted estimate for the fractal dimension, the similarity dimension, which is based on the concept of self-similarity. The similarity dimension, d_S, of a self-similar shape can be simply expressed as (see, e.g., [7])

$$N\ r^{d_S} = 1 \tag{2-11a}$$

or

$$d_S = \frac{\ell n\ N}{\ell n\ (1/r)} = -\frac{\ell n\ N}{\ell n\ r} \tag{2-11b}$$

where N is the number of similar parts within the whole, and r is the similarity or scaling ratio of an individual part to the whole. Mathematicians often resort to the similarity dimension to estimate the Hausdorff-Besicovitch or the fractal dimension because of its ease in computation [5].

For illustration, let us examine the standard self-similar shapes given in Figure 2-4, including a straight line of unit length, a square of unit area, a rectangle of area $A \cdot B$, with A being its length and B its width, a cube of unit volume, and a rectangular parallelepiped of volume $C \cdot D \cdot E$, with C being its length, D its width, and E its thickness. Since the Euclidean dimension of the straight line is one, for every integer "base" q, the "whole" interval may be paved by $N = q$ parts. Conversely, each part can be deduced from the whole by a similarity ratio of

$$r = \frac{1/q}{1} = \frac{1}{q} = \frac{1}{N} \tag{2-12}$$

This indicates that

$$N\ r^1 = 1 \tag{2-13}$$

Likewise, because the plane's Euclidean dimension is two, whatever the integer value of q, the whole surface of the square can be paved exactly by $N = q^2$ parts. Conversely, each part can be deduced from the whole by a similarity ratio of

$$r = \frac{1/q}{1} = \frac{1}{q} = \frac{1}{N^{1/2}} \tag{2-14}$$

thereby demonstrating that

$$N\ r^2 = 1 \tag{2-15}$$

The rectangle is divided into s similar parts so that it comprises $N = s$ parts. This is accomplished by imposing the following constraint:

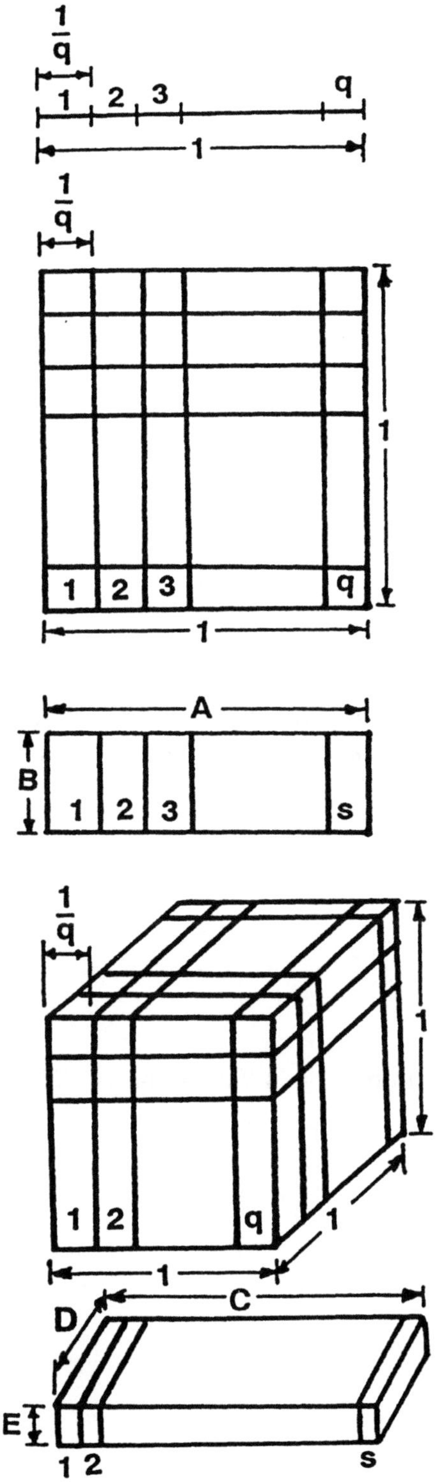

$$d_T = d_F = d_E = 1$$
$$N = q$$
$$r = \frac{1/q}{1} = \frac{1}{q} = \frac{1}{N}$$

$$d_T = d_F = d_E = 2$$
$$N = q^2$$
$$r = \frac{1/q}{1} = \frac{1}{q} = \frac{1}{N^{1/2}}$$

$$d_T = d_F = d_F = 2$$
$$N = s$$
$$r = \frac{1}{s^{1/2}} = \frac{1}{N^{1/2}}$$

$$d_T = d_F = d_E = 3$$
$$N = q^3$$
$$r = \frac{1/q}{1} = \frac{1}{q} = \frac{1}{N^{1/3}}$$

$$d_T = d_F = d_E = 3$$
$$N = s$$
$$r = \frac{1}{s^{1/3}} = \frac{1}{N^{1/3}}$$

Figure 2-4. "Standard" of self-similarity (see, e.g., [1]).

$$A : B = B : \frac{A}{s} \tag{2-16}$$

or

$$\frac{A}{B} = \frac{B}{A/s} \tag{2-17}$$

This leads to

$$B^2 = \frac{A^2}{s} \tag{2-18}$$

and thus [9],

$$r = \frac{B}{A} = \frac{A/s}{B} = \frac{1}{s^{1/2}} = \frac{1}{N^{1/2}} \tag{2-19}$$

or

$$N\ r^2 = 1 \tag{2-20}$$

The Euclidean dimension of the cube is 3. Obviously, its volume can be subdivided into $N = q^3$ parts, each of which is similar to the original cube. Thus, the similarity ratio, r, is obtained as

$$r = \frac{1/q}{1} = \frac{1}{q} = \frac{1}{N^{1/3}} \tag{2-21}$$

or

$$N\ r^3 = 1 \tag{2-22}$$

The rectangular parallelepiped is divided into $N = s$ similar parts. This can be accomplished by imposing the constraint that

$$C : D : E = D : E : \frac{C}{s}$$

or

$$\frac{C/D}{D/E} = \frac{D/E}{E/(C/s)} \tag{2-23}$$

This, in return, leads to

$$\frac{C}{D}\,\frac{E}{C/s} = \left(\frac{D}{E}\right)^2 \tag{2-24}$$

or

$$\left(\frac{E}{D}\right)^3 = \frac{1}{s} \tag{2-25}$$

Thus, the similarity ratio, r, is obtained as

$$r = \frac{E}{D} = \frac{1}{s^{1/3}} = \frac{1}{N^{1/3}}\ , \tag{2-26}$$

thereby demonstrating that

$$N\ r^3 = 1 \tag{2-27}$$

Note that Eqs. 2-11a and 2-11b may be regarded as the generalization or extension of the expressions given by Eqs. 2-13, 2-15, 2-20, 2-22 and 2-27 to objects with irregular but self-similar geometries.

It should be pointed out that for an object in a space with d_E of 3, its $(\text{length})^1$, $(\text{area})^{1/2}$, and $(\text{volume})^{1/3}$ all provide alternative measures of its linear extent, and the ratio of any two of them may be regarded as a shape parameter independent of the units of measurement. This observation is generally applicable to an object for which the topological dimension is equal to the Euclidean dimension. Prime examples include a line, an area and a volume whose Euclidean dimensions are one, two, and three, respectively, and their topological dimensions are identical to the respective Euclidean dimensions. The geometries of such objects are considered "standard" [10]. However, the fractal dimension of the rugged perimeter of a flat and smooth area, whose d_T is 1 and d_E is 2, lies between 1 and 2, and that of the rugged surface of a uniform or homogeneous volume, whose d_T is 2 and d_E is 3, lies between 2 and 3. The perimeter of the former is "non-standard" or fractal, but its area is standard; similarly, the surface of the latter is "non-standard" or fractal, but its volume is standard. In other words, some objects are standard as well as fractal. A concrete example of such objects is a mammalian brain; its irregular surface is non-standard or fractal, but its volume is essentially standard. Conceptually, the fractal dimension of the nonstandard geometry of an object can be estimated by relating it to the Euclidean dimension of its standard geometry [10].

Let the G-length be the length measured with a yardstick of length G, and the G-area be the area measured in units of G^2. For an object with an Euclidean dimension, d_E, of 2, the dependence of G-length upon G is generally non-standard because of the ruggedness of a linear dimension characterizing this object, e.g., its perimeter; in contrast, the dependence of G-area on a flat or smooth plane on G^2 is standard. Thus, it is highly plausible that the G-length is related to the G-area as

$$(\text{G-length})^{1/d_F} \propto (\text{G-area})^{1/2} \tag{2-28}$$

Similarly, for an object with d_E of 3, the G-area (non-standard) is related to the G-volume (standard) as

$$(\text{G-area})^{1/d_F} \propto (\text{G-volume})^{1/3} \tag{2-29}$$

The fractal dimensions evaluated from these two relationships are indicative of irregularity of lines or surfaces [10].

Increasingly irregular lines on a flat surface or plane given in Figure 2-5 demonstrate qualitatively or approximately how the fractal dimension, d_F, increases with the increase in the ruggedness of the lines, whereas their topological dimensions, d_T, remain invariant at 1, and their Euclidean dimensions, d_E, shift abruptly from 1 to 2 at the outset [11].

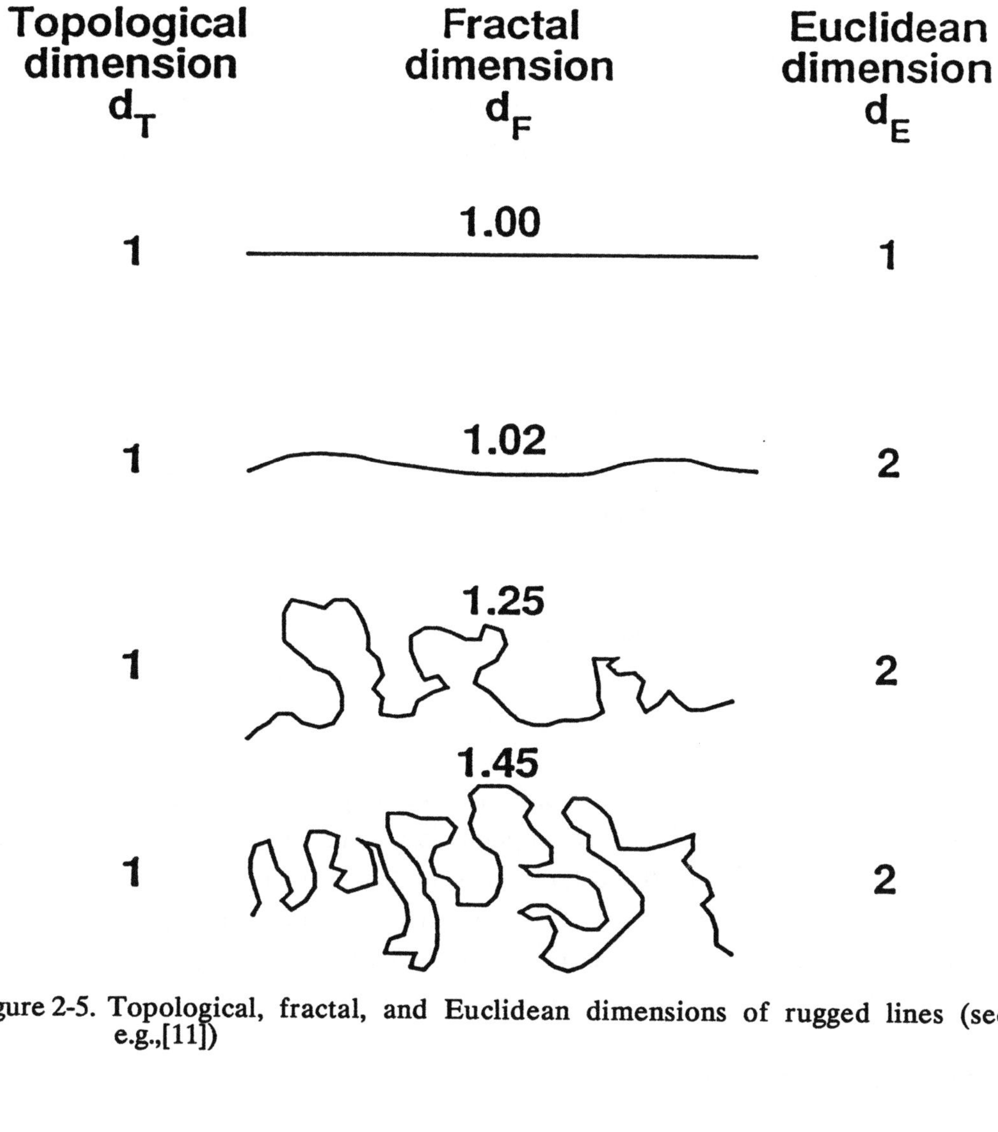

Figure 2-5. Topological, fractal, and Euclidean dimensions of rugged lines (see, e.g.,[11])

3. EXAMPLES OF FRACTAL GEOMETRY

Examples of fractal geometry abound. As mentioned earlier, the fractal dimension, d_F, is constrained by the topological dimension, d_T, from below and the Euclidean dimension, d_E, from above. The fractal dimension of a rugged line on the plane with an Euclidean dimension of 2 is between 1 and 2, as demonstrated in Figure 2-5. Similarly, the fractal dimension of a rugged line is between 1 and 3 in the space with an Euclidean dimension of 3. Relatively simple examples include, among others, the Cantor set whose topological and Euclidean dimensions are 0 and 1, respectively, and $0 < d_F < 1$, and the irregular surface of a particulate object, e.g., a catalyst, whose topological and Euclidean dimensions are 2 and 3, respectively, and $2 < d_F \leq 3$.

3.1 Cantor Set ($0 < d_F < 1$)

This well-known group of examples of geometry, having fractal dimensions between 0 and 1, is also sometimes termed the Cantor dust; it is named after the nineteenth-century mathematician Georg Cantor (see, e.g., [5]). Informally, the term "dust" is equivalent to a set of disconnected points with the set having a topological dimension, d_T, of 0.

One way to generate a Cantor set is by the procedure illustrated in Figure 3-1, in which a straight line of unit length is initially given ($k = 0$). To initiate the procedure, its middle one-third is removed ($k = 1$); this procedure is repeated for the remaining segments ($k = 2, 3, 4, 5, \cdots$). This eventually yields the dust of remaining points ($k \longrightarrow \infty$), constituting a Cantor set. These points are infinitely many, but their total length is negligible. Since the Cantor set is a set of points formed on a straight line, its topological and Euclidean dimensions are 0 and 1, respectively. The Cantor set is also termed as the Cantor discontinuum or fractal dust (see, e.g., [5]). It is noted in Figure 3-1 that the fractal dimension, d_F, of the Cantor set is 0.6309 which is between its topological dimension, d_T, of 0 and its Euclidean dimension, d_E, of 1.

3.2 Rugged Lines ($1 < d_F \leq 2$)

Methods of generating self-similar shapes are delineated here with the example of a Koch curve; this curve is a geometrically symmetric but monstrous shape with characteristic self-similarity embedded in it. It is named after a Swedish mathematician von Koch in the 1900's; its fractal dimension, d_F, lies between d_T of 1 and d_E of 2 (1 <

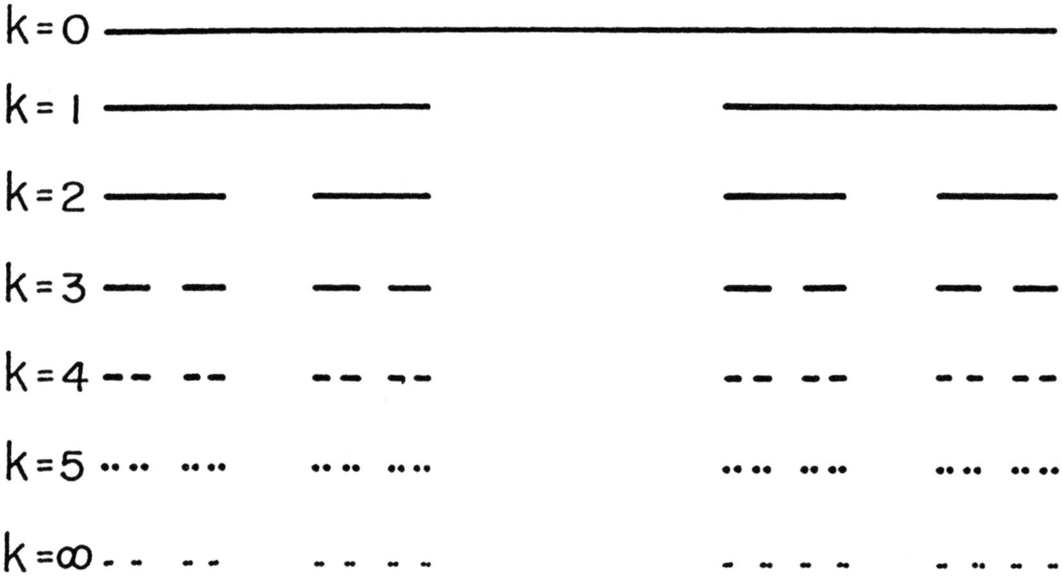

Figure 3-1. Generation of the Cantor set (see, e.g., [5]):

$$N = 2$$

$$r = \frac{1/3}{1} = \frac{1}{3}$$

$$Nr^{d_F} = 2 \left[\frac{1}{3} \right]^{d_F} = 1$$

$$d_F = \frac{\ell n\ 2}{\ell n\ 3} = 0.6309.$$

$d_F \leq 2$) (see, e.g., [5]). The presence of multi-fractal dimensions in a rugged profile is also discussed in this subsection.

3.2.1 Koch Curve and Lake

The generator of these curves is presented in Figure 3-2. A generator is the fundamental geometry giving rise to a complex pattern. The ℓ in the figure is obtained by dividing a line of unit length into three equal parts, thus yielding the generator comprising four segments of ℓ's. Hence, the fractal dimension, d_F, of the Koch curve is equal to 1.26, as demonstrated in the caption of the figure.

A triadic Koch lake in Figure 3-3 is composed of three Koch curves. Its construction begins with an equilateral triangle, with each side of length one (k = 0 in Figure 3-4). It is replaced by a generator (k = 1); each straight line segment of the generator is then replaced by a generator one-third the size and so on (k = 2, 3, · · ·). Figure 3-5 demonstrates in detail the steps for generating a triadic Koch lake through the evolution of a Koch curve on each side of the lake.

The area enclosed by a circle drawn around the original triangle (indicated by dotted lines) is shown in Figure 3-3. Obviously, this area is finite; however, the length of the perimeter of the Koch lake is

$$L_p = \lim_{k \to \infty} 3\left(\frac{4}{3}\right)^k \longrightarrow \infty, \qquad (3\text{-}1)$$

thereby indicating that an infinitely long line surrounds a finite area.

3.2.2 Fractal Dimension of a Rugged Profile

As has been illustrated by the Koch Lake, a monstrous shape can be fractal and self-similar. Although the shapes of natural objects, such as coastlines, boundaries between countries, and profiles of fine particles, are not always precisely self-similar, they can often be regarded approximately as fractals.

According to Richardson [13], two neighboring countries may measure the length of their shared boundary differently. For example, Spain considers the length of its shared boundary with Portugal to be 987 km, while Portugal regards it to be 1214 km. Figure 3-6 presents lengths of various curves measured with different scales of the measuring unit, λ; this figure is known as the Richardson plot. The length of a circle stabilizes very rapidly near a well-determined value as λ becomes smaller. The lengths

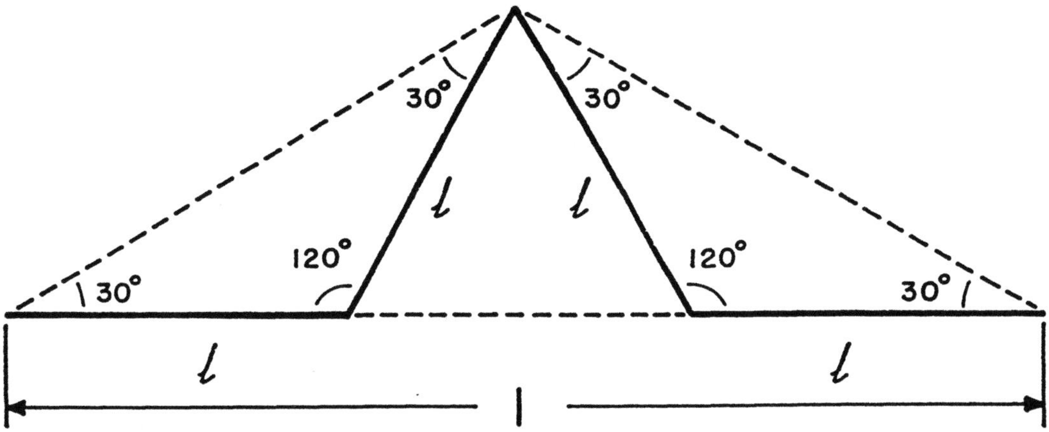

Figure 3-2. Generator of a Koch curve:

$$N = 4$$

$$r = \frac{\ell}{1} = \frac{1}{3}$$

$$Nr^{d_F} = 4\left(\frac{1}{3}\right)^{d_F} = 1$$

$$d_F = \frac{\ell n\ 4}{\ell n\ 3} = 1.26.$$

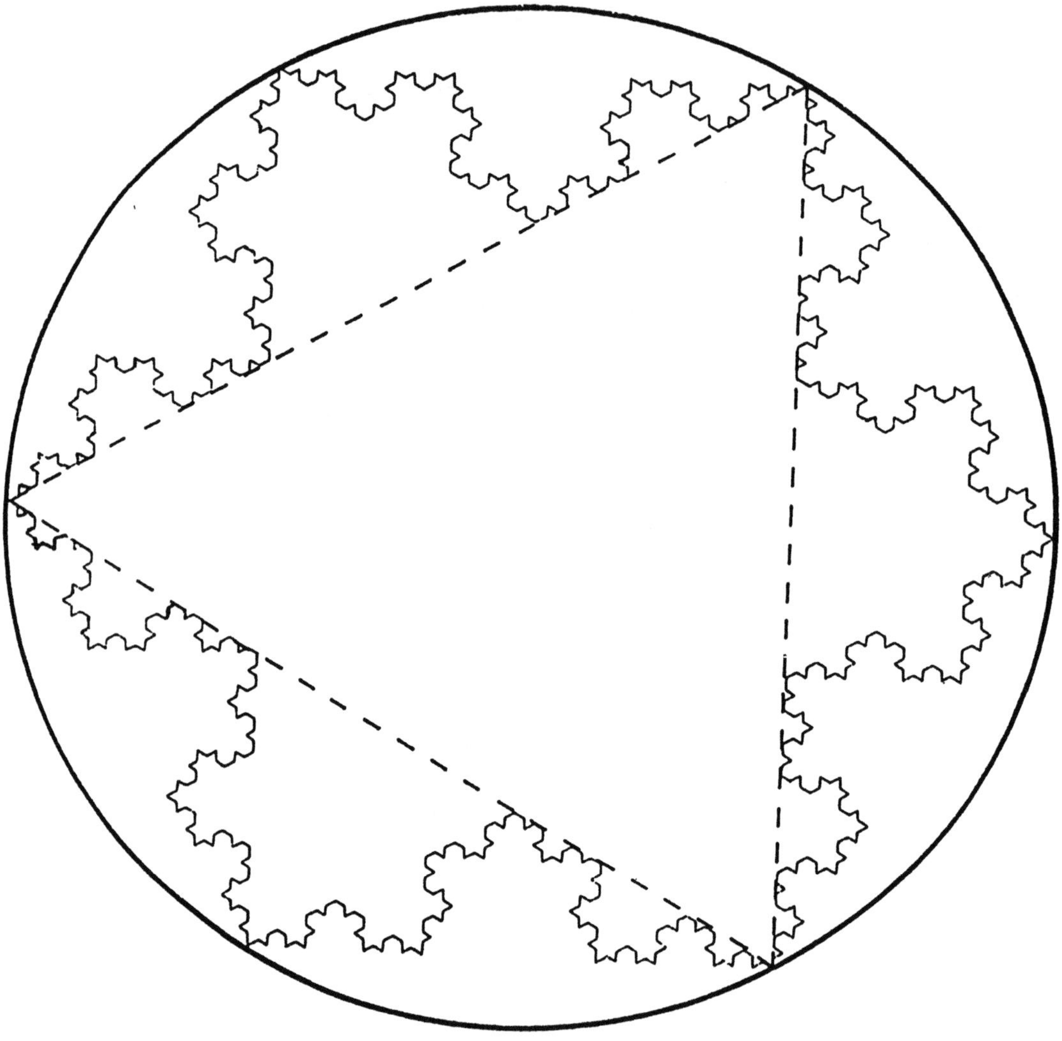

Figure 3-3. Triadic Koch lake [5].

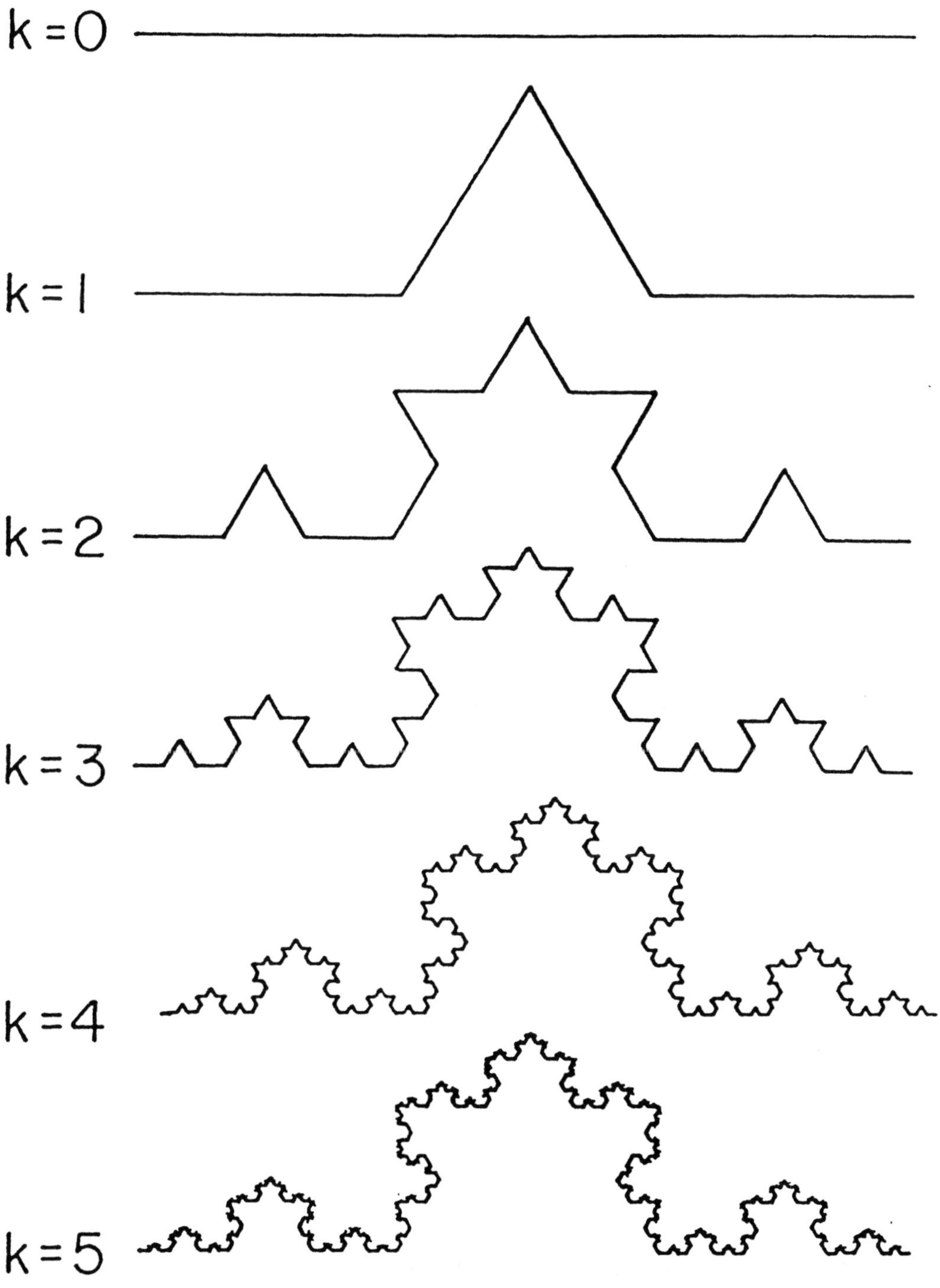

Figure 3-4. Procedure for generating a Koch curve: k is the step number (see, e.g., [12]).

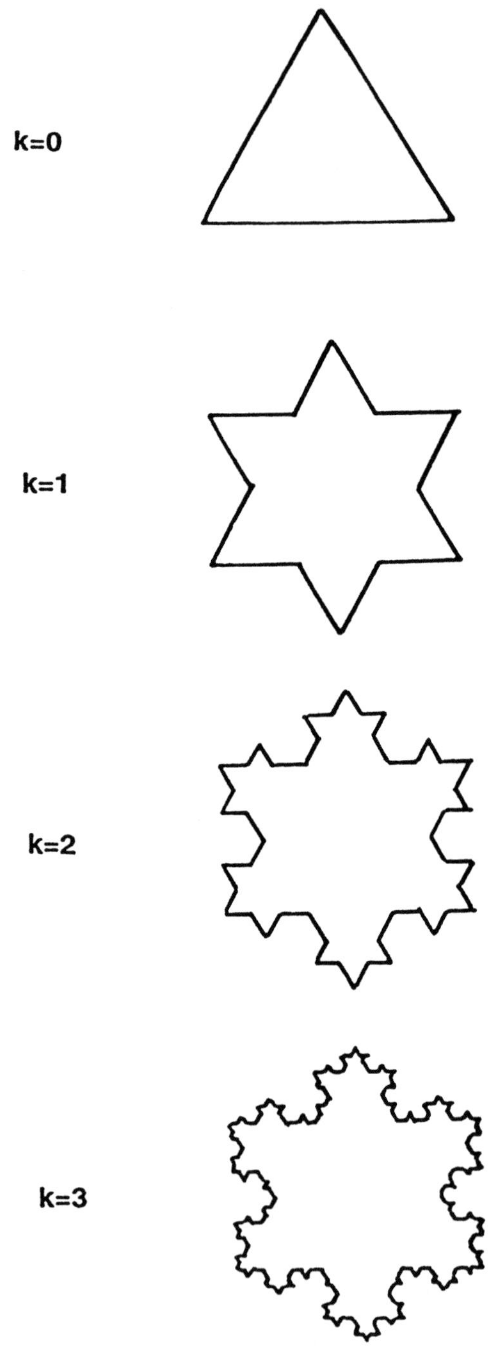

Figure 3-5. Procedure for constructing a Triadic Koch lake: k is the step number (see, e.g., [5]).

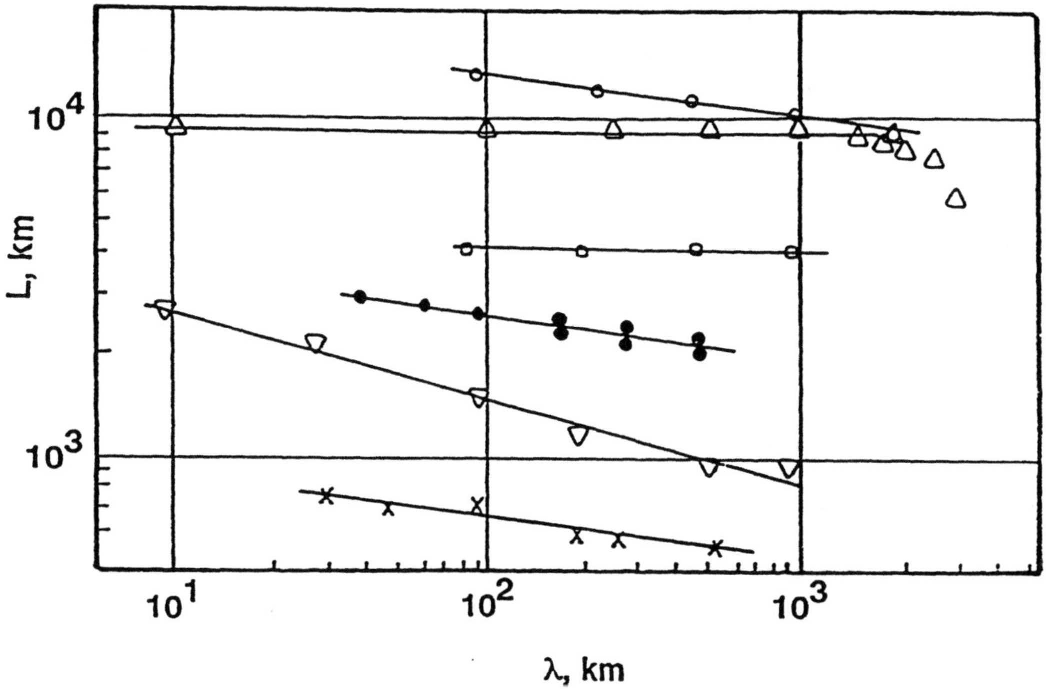

Figure 3-6. Richardson plot [5]:

- o Australian coast
- Δ Circle
- □ South African coast
- • German land-frontier
- ▽ West coast of Britain
- × Land-frontier of Portugal.

of all fractal curves, the so-called nonstandard curves, increase with the decreasing length of the measuring unit, λ, yielding the following expression;

$$L(\lambda) = F\,\lambda^{a} \tag{3-2}$$

where F is a proportional constant. While Richardson has not rationally interpreted the exponent, a, Mandelbrot [5] has related it to the fractal dimension, d_F, as follows:

$$a = -d_F+1 \tag{3-3}$$

This relationship may be obtained from a defining equation of self-similarity as shown below.

Equation 2-11a, with d_S interpreted as d_F, can be rewritten as

$$N\,r\,(r)^{d_F-1} = 1 \tag{3-4}$$

or

$$N\,r = r^{-d_F+1} \tag{3-5}$$

Since

$$\lambda \propto r \tag{3-6}$$

and

$$L(\lambda) \propto N\,r, \tag{3-7}$$

we see that

$$L(\lambda) \propto \lambda^{-d_F+1} \tag{3-8}$$

The Richardson plot is applied to the description of rugged fine particles and fine particle systems such as fluidized beds and packed powder beds. Kaye *et al.* [14] have presented the method of evaluating the fractal dimensions of the profiles of fine particles. Their approach, depicted in Figure 3-7, is similar to that adopted by Richardson for characterizing the structured walk technique. To estimate the perimeter of a fine particle profile according to the structured walk technique, we stride around the profile with a compass having a step size of λ. This yields a polygon as indicated by the dotted line segments in Figure 3-7. The length of the final fractional length for completing the polygon is $\varsigma\lambda$ as indicated; obviously, $\varsigma < 1$. The estimate of the perimeter based on the step size, λ, is $(n+\varsigma)\lambda$, in which n is an integer. If the fine particle profile has a fractal boundary, a logarithmic plot of the perimeter estimated against the step size yields a straight line with a slope of $-d_F+1$; d_F is the fractal

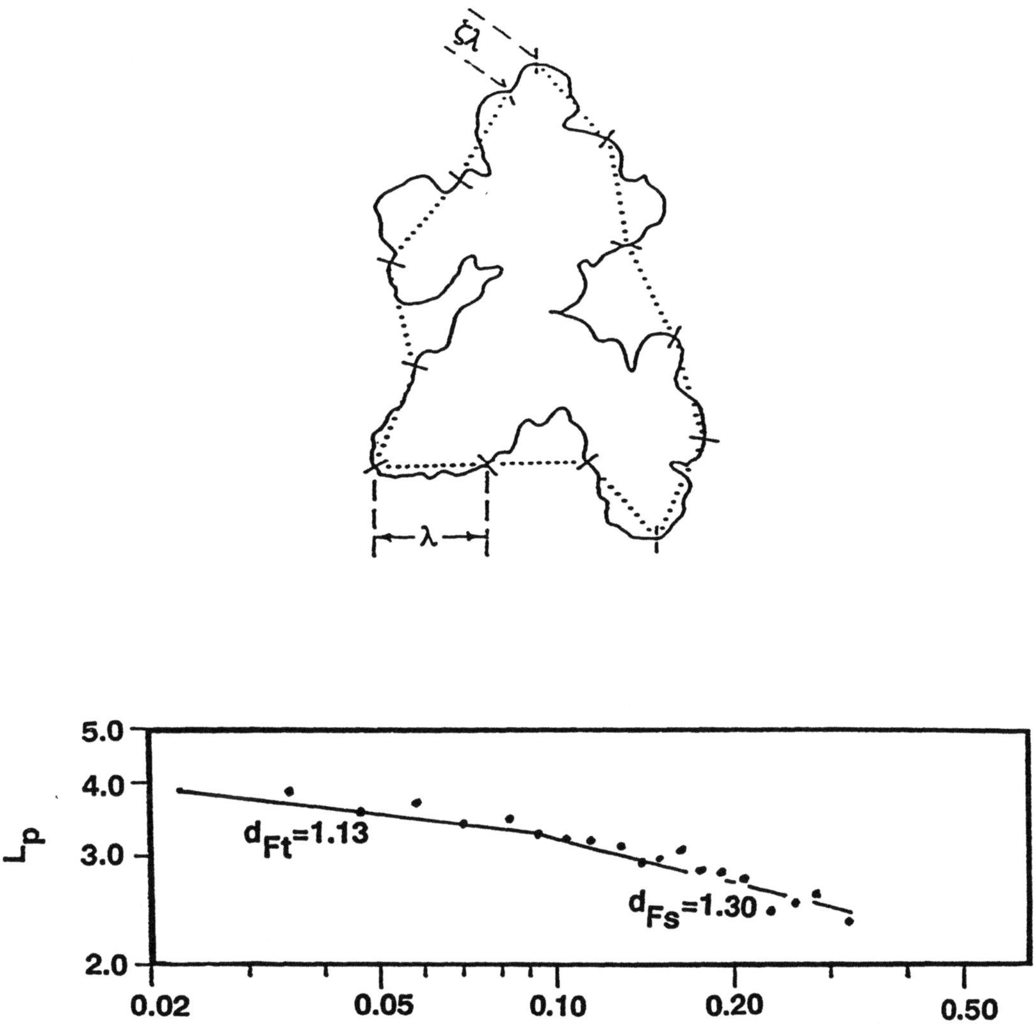

Figure 3-7. Structured walk technique (see, e.g., [14]):

d_{Fs} = structural fractal dimension
d_{Ft} = textural fractal dimension.

dimension of the boundary.

Other methods, e.g., the signature wave form technique illustrated in Figure 3-8, are also available for evaluating shapes of fine particles. The signature wave form technique transforms the profile into a wave form by measuring the magnitude of the vector, $\vec{R}$, as a function of the angle, θ ($0 \leq \theta < 360$), rotated around the perimeter of the profile. This wave form function is then subjected to the Fourier analysis to decompose the boundary into its components comprising simple harmonic functions. Nevertheless, this method has the following disadvantages [15].

a. The value of R depends on the angular resolution (Figure 3-9a).
b. The value of R is not unique if it crosses a re-entrant feature of the profile (Figure 3-9b).
c. The presence of the sharp edge in the profile gives very high frequencies in the subsequent Fourier analysis; it is difficult to interpret the Fourier coefficient of the signature wave forms of such profiles.

3.2.3 Multi-fractal Dimension

In practice, we come across various fractal dimensions when the perimeter of a particle profile is scrutinized at different levels of resolution. This is evident in Figure 3-7, where the particle profile constitutes two fractal dimensions, namely,

$$d_{Ft} = 1.13$$

and

$$d_{Fs} = 1.30$$

at different levels of scrutiny. The former is called the "textural" fractal dimension and the latter the "structural" fractal dimension. The texture is the fine scale structure, and the structure is the overall gross structure of the profile; however, the dichotomy between them is not crisp, i.e., it is fuzzy [17].

Most fine particles have more than one fractal dimension; in fact, Kaye [17] has stated that the use of different levels of scrutiny or resolution in measuring the geometry of a natural object can give rise to different fractal dimensions. For example, let us consider a rugged island, and suppose that all coastal indents smaller than 10m have been filled with concrete in which boulders, each with a diameter of 1m, have been lodged (Figure 3-10). Then, the perimeter of the island is measured with the stride

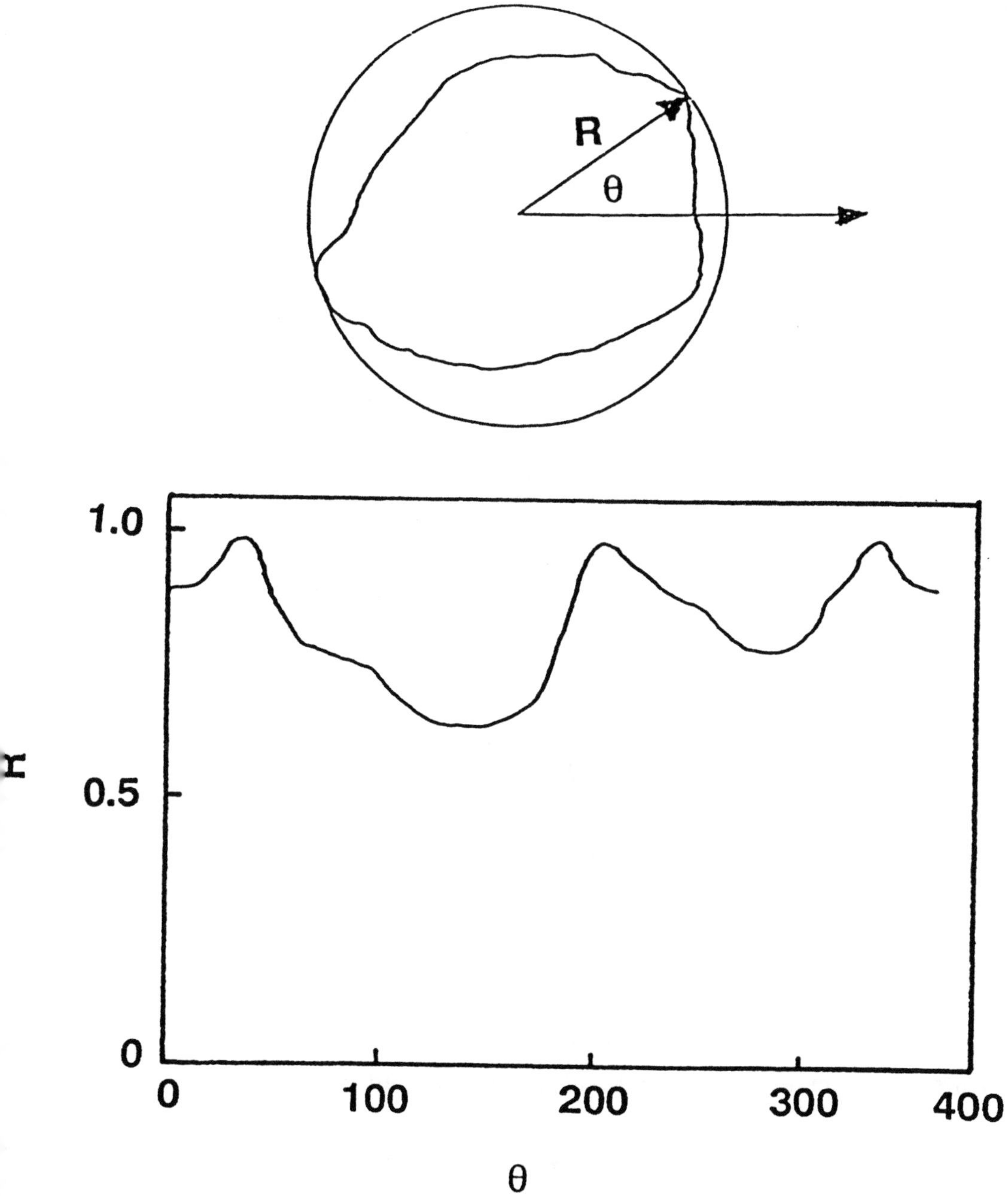

Figure 3-8. Signature wave form technique [16].

Reprinted by permission of Elsevier Sequoia S.A.

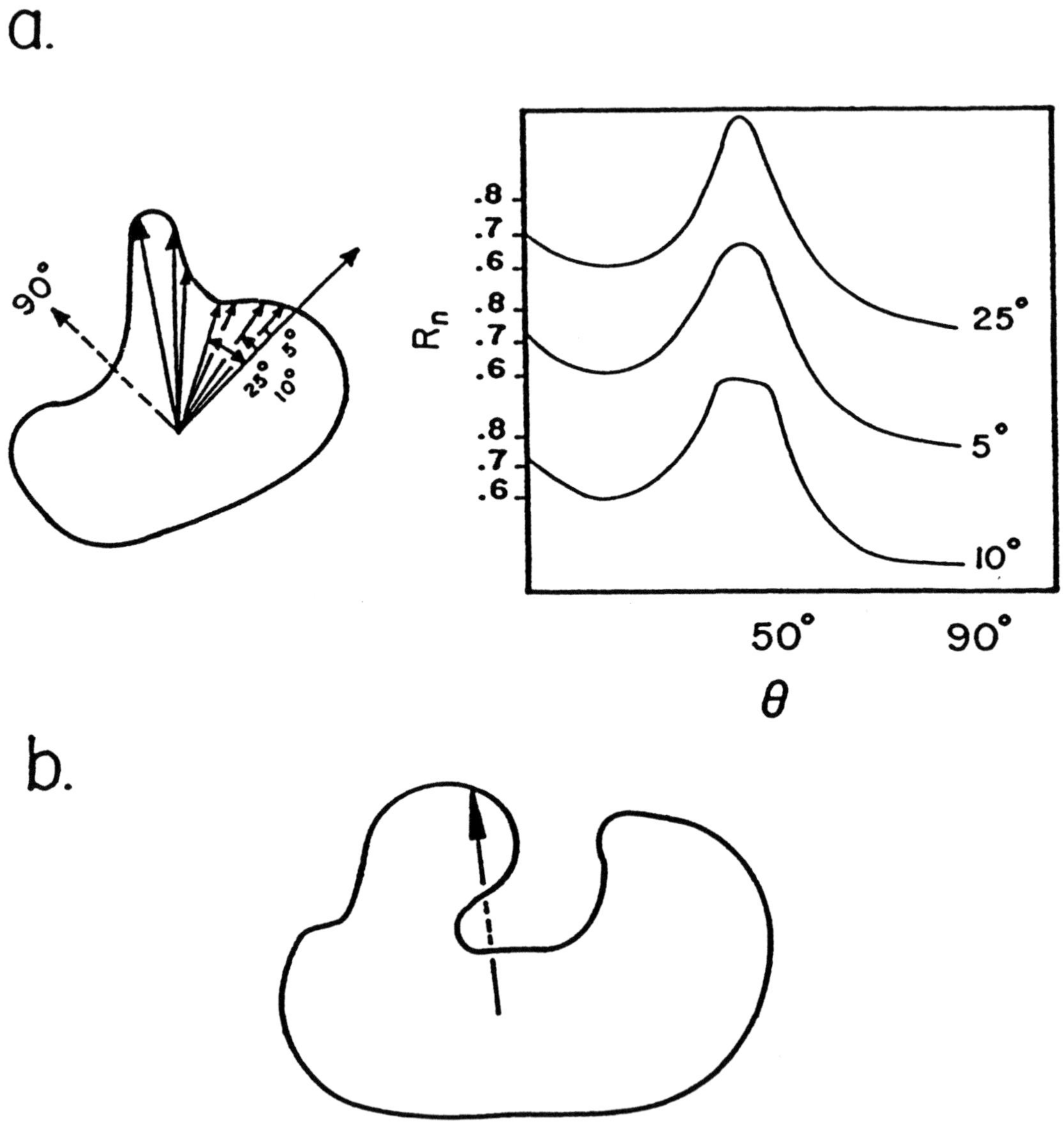

Figure 3-9. Demonstration of difficulties encountered by the signature wave form technique [16].

Reprinted by permission of Elsevier Sequoia S.A.

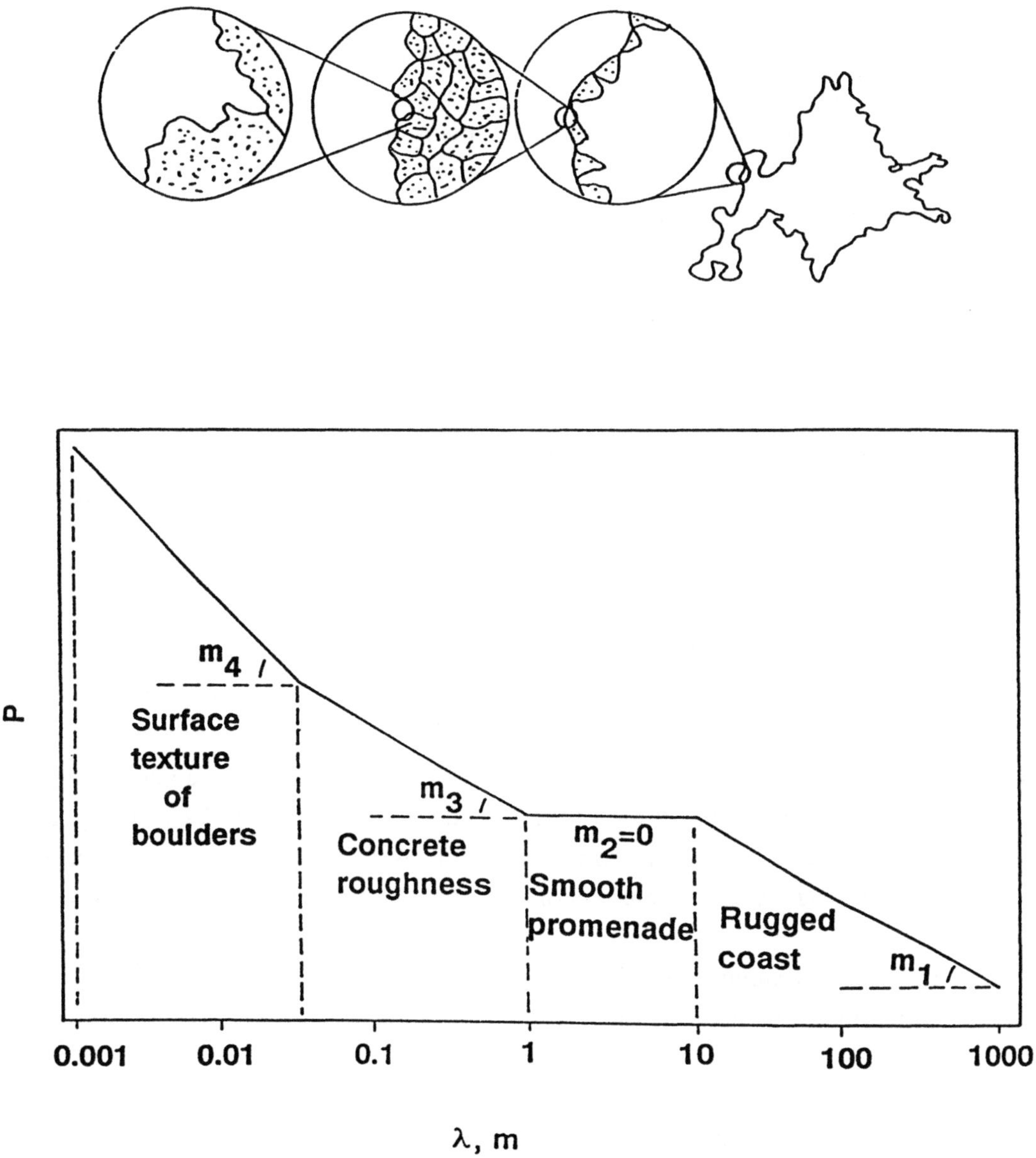

Figure 3-10. Multi-fractal dimensions [17].

length, λ, ranging from 0.001m to 1000m. This is followed by the construction of a Richardson plot with the resultant measurements. When λ is greater than 10m a rugged coast is identified, the island becomes a polygon composed of the straight lines when λ is greater than 1m and less than 10m. Eventually, the concrete roughness is measured when λ is less than 1m, and finally, the surface texture of boulders is sensed if λ becomes sufficiently small. Therefore, different levels of the structural details in terms of its perimeter can be described with the multi-fractal dimension (Figure 3-10).

Application. For a realistic example of how to compute fractal dimensions of particles, readers are referred to Appendix 1 (also see [18]). This Appendix describes, with considerable details, the computational techniques for extracting perimeter fractal dimensions of char particles.

3.3 Irregular Surface ($2 < d_F \leq 3$)

Irregularity of surfaces plays an important role in heterogeneous chemical reaction, adsorption and other phenomena and processes involving interactions between solid surfaces and fluid, and even among different solid particles. Since the profile of an irregular surface is only its partial expression, surface science tends to describe the geometrically irregular surfaces by treating ruggedness as a deviation from standard or regular shapes. Pfeifer and Avnir [19] have proposed that the fractal approach analogous to the one dealing with a profile of an irregular surface can be applied to the analysis of the geometry of a rugged surface, if it is characteristically self-similar at various levels of resolution. For such a surface, the degree of irregularity can be defined by one number, i.e., the fractal dimension, $2 < d_F \leq 3$, where 2 is its topological dimension and 3 its Euclidean dimension. Pfeifer and Avnir [19] have proposed three methods for evaluating fractal dimensions of the irregular surfaces; these methods are based on:

i. formation of monolayers of different adsorbates,
ii. formation of monolayers on adsorbent particles of different sizes but similar shapes,

and

iii. determination of the pore-size distribution.

3.3.1 Monolayers of Different Adsorbates

In a 3-dimensional space, the area of an irregular surface can be estimated with flat squares as illustrated in Figure 3-11. Suppose that the flat squares are replaced by balls, each with a radius of b, under the assumption that only one ball can be located on each square as shown in Figure 3-12. The number of balls covering the surface, N_b, increases with a decrease in the radius of the ball, b. If the surface is flat, the rate of this increase is inversely proportional to the square of b. When the surface is fractal, i.e., its geometrical feature is self-similar, from the definition of the fractal (similarity) dimension, d_F, given in Eq. 2-11a or 2-11b with d_S replaced by d_F, we have [19]

$$N_b \propto b^{-d_F}, \qquad 2 < d_F \leq 3 \tag{3-9}$$

The fractal dimension of the surface, d_F, can be evaluated with adsorbate molecules of various sizes but similar shapes, e.g., Helium, Nitrogen, Argone, Cripton and Carbon Dioxide on a given adsorbent as illustrated in Figure 3-13. To estimate the surface area of a solid object with a typical adsorption experiment, the number of molecules adsorbed on a unit volume or mass of the adsorbent, n, corresponding to monolayer formation, is determined. Thus,

$$n \propto \frac{N_b}{R^3} \tag{3-10a}$$

where R is the characteristic length of the adsorbent particle, which is constant. Equation 3-9, therefore, can be expressed as follows:

$$n \propto b^{-d_F}, \qquad b_{min} \leq b \leq b_{max} \tag{3-10b}$$

where b_{min} is the smallest adsorbate radius and b_{max} the largest.

In the preceding paragraph, it is assumed that the surface is covered by spherical adsorbate balls with a radius of b. However, the adsorbate molecules need not be spherical. If they belong to a homologous series, e.g., long-chain n-alkane and nitrogen, the ratio of the square of the linear extent, b^2, to the effective cross-sectional area, σ, remains invariant for all members of the homologous series [20], i.e.

$$b^2 \propto \sigma$$

or

$$b_0 \propto \sigma^{1/2}$$

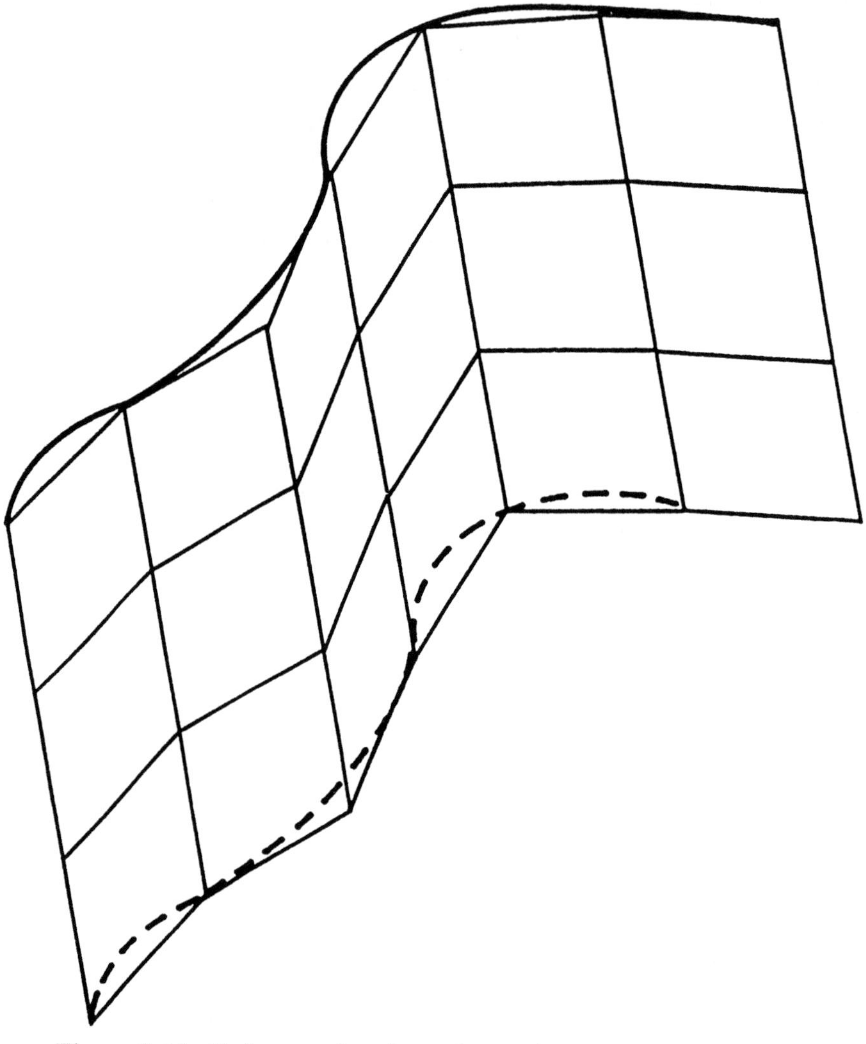

Figure 3-11. Estimate of an irregular surface with flat squares.

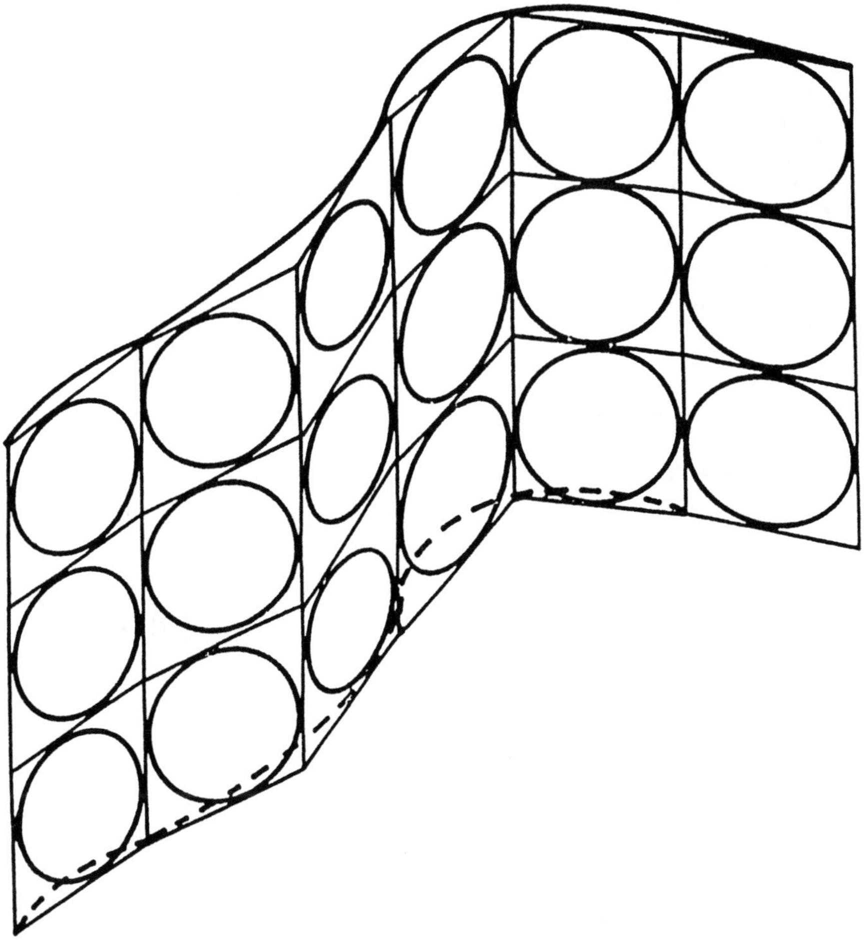

Figure 3-12. Estimate of an irregular surface with balls.

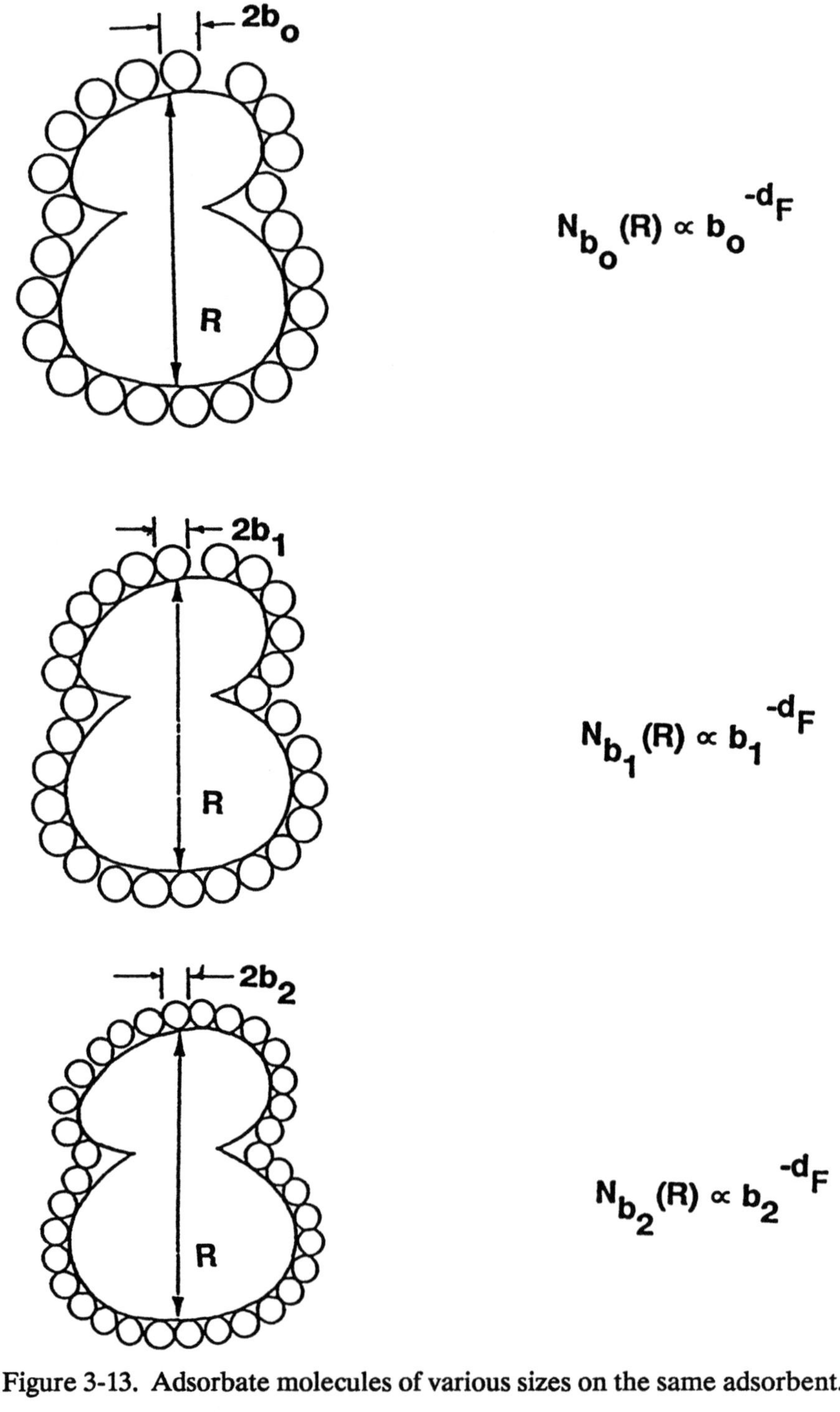

Figure 3-13. Adsorbate molecules of various sizes on the same adsorbent.

In this case, Eq. 3-10b becomes

$$n \propto \sigma^{-d_F/2} \qquad (3\text{-}11)$$

Since

$$A_s = n\sigma \qquad (3\text{-}12)$$

where A_s is the surface area of the adsorbent, we have

$$A_s \propto \sigma^{-d_F/2}\sigma \propto \sigma^{-(d_F-2)/2} \qquad (3\text{-}13)$$

Therefore, selecting the series of adsorbates to be isotropic or homologous ensures that a well-defined functional relationship exists between $\ell n\, \sigma$ and ℓn A or ℓn n. According to Eqs. 3-11 and 3-13, a plot of ℓn n against $\ell n\, \sigma$ and that of ℓn A against $\ell n\, \sigma$ produce straight lines with slopes equal to $-d_F/2$ and $-(d_F - 2)/2$, respectively. Since

$$2 < d_F \leq 3$$

for a fractal surface, the natural boundaries of the former are

$$-\frac{3}{2} \leq -\frac{d_F}{2} < -1$$

and those of the latter are

$$-\frac{1}{2} \leq -\frac{d_F-2}{2} < 0$$

3.3.2 Monolayers on Adsorbent Particles of Different Sizes

Equations 3-10a and 3-10b imply that the number of adsorbate molecules on the adsorbent surface having a characteristic length of R_0, N_{b_0} is functionally related to the radius of adsorbate molecules, b_0, as

$$N_{b_0}(R_0) \propto b_0^{-d_F}, \qquad b_{min} \leq b_0 \leq b_{max} \qquad (3\text{-}14)$$

Similarly, when the radius of the adsorbate is b_1 or b_2, we may write as (Figure 3-14)

$$N_{b_1}(R_0) \propto b_1^{-d_F}, \qquad b_{min} \leq b_1 \leq b_{max} \qquad (3\text{-}15a)$$

or

$$N_{b_2}(R_0) \propto b_2^{-d_F}, \qquad b_{min} \le b_2 \le b_{max} \tag{3-15b}$$

In these expressions, N_{b_1} and N_{b_2} are the numbers of adsorbate molecules on the adsorbent surface with radii of b_1 and b_2, respectively.

Now, let us visualize that two larger adsorbent particles, one with the characteristic length of R_1 and the other R_2, are covered with the same adsorbate molecule with the radius, b_0, as indicated in Figure 3-14. Furthermore, suppose that the shapes of these two larger absorbent particles are similar to that of the original adsorbent particle, thereby having the same fractal dimension and suppose that the following conditions are satisfied:

$$\frac{R_1}{R_0} = \frac{b_0}{b_1} \tag{3-16a}$$

and

$$\frac{R_2}{R_0} = \frac{b_0}{b_2} \tag{3-16b}$$

The number of adsorbate molecules with a radius of b_0 on the adsorbent particle, having a characteristic length of R_1, is denoted by $N_{b_0}(R_1)$. Similarly, the number of adsorbate molecules with a radius of b_1 on the adsorbent particle, having a characteristic length of R_0, is denoted by $N_{b_1}(R_0)$. Since these two numbers are identical because of Eqs. 3-16a and 3-16b, we have (Figure 3-14)

$$N_{b_0}(R_1) = N_{b_1}(R_0) \tag{3-17}$$

From Eqs. 3-15a and 3-16a,

$$N_{b_0}(R_1) \propto (R_0 b_0)^{-d_F} R_1^{d_F} \tag{3-18}$$

Similarly, from Eqs. 3-15b and 3-16b,

$$N_{b_0}(R_2) \propto (R_0 b_0)^{-d_F} R_2^{d_F} \tag{3-19}$$

Note that the number of identical adsorbate molecules with a radius of b on various adsorbent particles, each with a characteristic length of R, can generally be written as

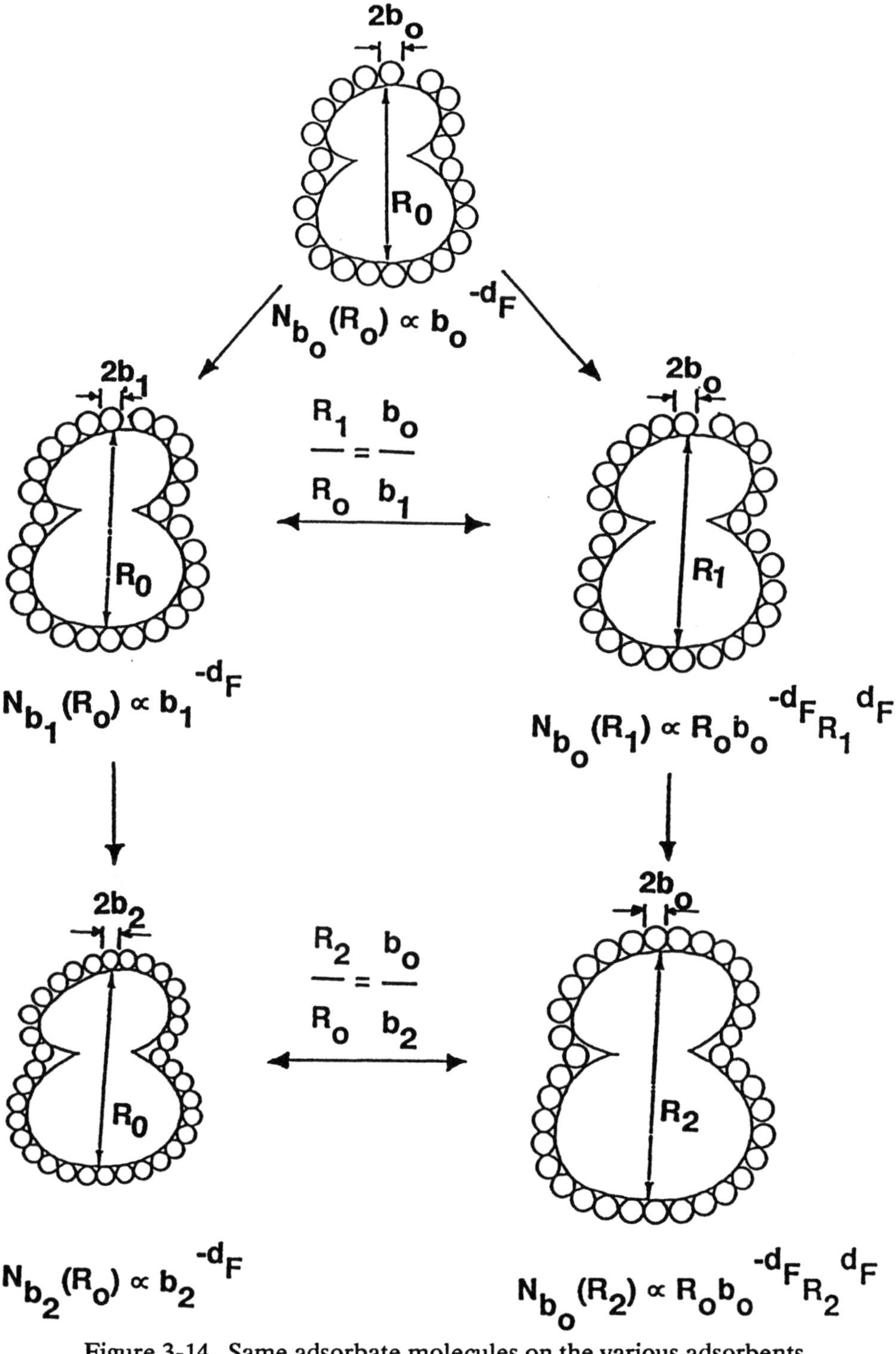

Figure 3-14. Same adsorbate molecules on the various adsorbents.

$$N_b(R) \propto R^{d_F} \tag{3-20}$$

and the number of molecules per unit mass of the adsorbent is

$$n \propto \frac{N_b(R)}{R^3} \tag{3-21}$$

Thus, similar to Eq. 3-10, Eq. 3-20 can be rewritten as

$$n \propto R^{d_F - 3} \tag{3-22}$$

Equation 3-14 is applicable only to the adsorbate molecules with their radii between b_{min} and b_{max}. Similarly, the characteristic length of the adsorbent, R, expressed in Eqs. 3-20 and 3-22 also has an applicable range. This range is

$$R_{min} \le R \le R_{max} \tag{3-23}$$

where

$$R_{min} = \frac{b_0}{b_{max}} R_0$$

and

$$R_{max} = \frac{b_0}{b_{min}} R_0$$

To recover the maximum information on the irregularity of the surface, the smallest possible b, i.e., b_{min}, should be selected. If b is large, the self-similarity below this size can only be speculated [19].

3.3.3 Pore-size Distribution

The mercury porosimeter does not directly determine the surface area; it measures the size and number of pores in porous materials. The amount of mercury expelled from pores as a function of decreasing pressure gives information about the shape and structure of the pores (see, e.g., [21]). The physical basis of this device can be appreciated from the simple system illustrated in Figure 3-15. The force balance between the surface tension and the applied pressure on the mercury surface at the pore opening

$$\pi\, r^{*2} P = 2\pi\, r^{*} \eta \cos\phi$$

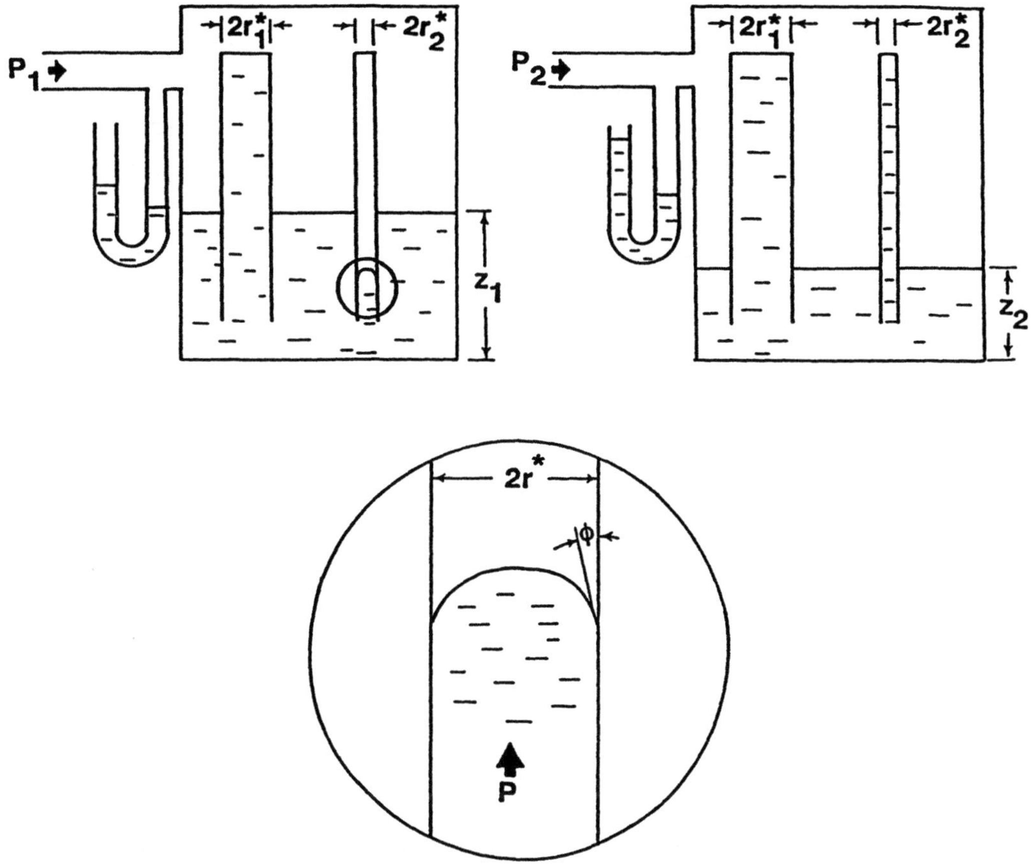

Figure 3-15. Schematic diagram of the mercury porosimeter:

P_1, P_2: applied pressures,

$P_1 < P_2$

r_1^*, r_2^*: radii of the cylinders

z_1, z_2: heights of mercury, $z_1 > z_2$.

gives rise to the Washburn equation (see, e.g., [21])

$$P = \frac{2\eta \cos\phi}{r^*} \tag{3-24}$$

where P is the applied pressure, r^* the radius of the pore, η the surface tension of mercury, and ϕ the contact angle made by the mercury with the inner surface of the pore (see, e.g., [15]). Each increment of the applied pressure causes the next smaller group of pores to be filled, with a concomitant increase in the total volume of mercury penetrated into the porous solid.

Suppose that the pores are cylindrical with radii of r^*'s and lengths of $h(r^*)$'s. Let V be the cumulative volume of the pores of radii greater than r_0^*, i.e.,

$$V(r_0^*) = \int_{r_0^*}^{\infty} m(r^*)\ [\pi r^{*2} h(r^*)]\ dr^* \tag{3-25a}$$

or

$$V(r^*) = \int_{r^*}^{\infty} f(x)\ dx = F(\infty) - F(r^*) \tag{3-25b}$$

where

$$f(x) = m(x)\ [\pi x^2 h(x)],$$

$$F(x) = \int f(x) dx$$

and $m(r^*)$ is the density function of the number of pores, i.e., $m(r^*)dr^*$ is the number of pores with radii between r^* and $r^* + dr^*$. Also, we have

$$\frac{dV(r^*)}{dr^*} = \frac{d}{dr^*}[F(\infty) - F(r^*)]$$

$$= -\frac{d}{dr^*}F(r^*)$$

$$= -m(r^*)\ [\pi r^{*2} h(r^*)]$$

Then, the number of the pores of radii between r^* and $r^* + dr^*$ is

$$m(r^*)dr^* = -\frac{dV}{dr^*}[\pi r^{*2} h(r^*)]^{-1} dr^* \tag{3-26}$$

Hence, the surface area of the pores of radii between r^* and $r^* + dr^*$ is written as

$$dA_s = \left\{ - \frac{dV}{dr^*} [\pi r^{*2} h(r^*)]^{-1} dr^* \right\} 2\pi r^* h(r^*)$$

$$= - \frac{dV}{dr^*} \frac{2}{r^*} dr^* \tag{3-27}$$

In an adsorption experiment, a pore of radius $r^* \geq r_0^* = b_0$ can accept the adsorbate molecule. Hence, the total surface area covered by the adsorbate molecules is

$$A_s = \int_{b_o}^{\infty} dA_s \tag{3-28}$$

and the total number of adsorbate molecules is

$$N_b \propto \frac{A_s}{\pi b^2} = \int_{b_o}^{\infty} \frac{dA_s}{\pi b^2} \tag{3-29}$$

or, because of Eq. 3-27,

$$N_b \propto \frac{1}{b^2} \int_{b_o}^{\infty} [- \frac{dV}{dr^*} \frac{2}{r^*}] dr^* \tag{3-30}$$

This right-hand side of this expression can be evaluated for pore-size distributions illustrated in Figure 3-16. Note that the following equation is applicable to the decreasing segment of each of these distribution, i.e., the portion of the line immediately after the peak:

$$- \frac{dV}{dr^*} \propto (r^*)^{-k}, \; k > 0 \tag{3-31}$$

Thus, Eq. 3-30 becomes

$$N_b \propto \frac{1}{b^2} \int_{b}^{\infty} (r^*)^{-k-1} dr^*$$

or

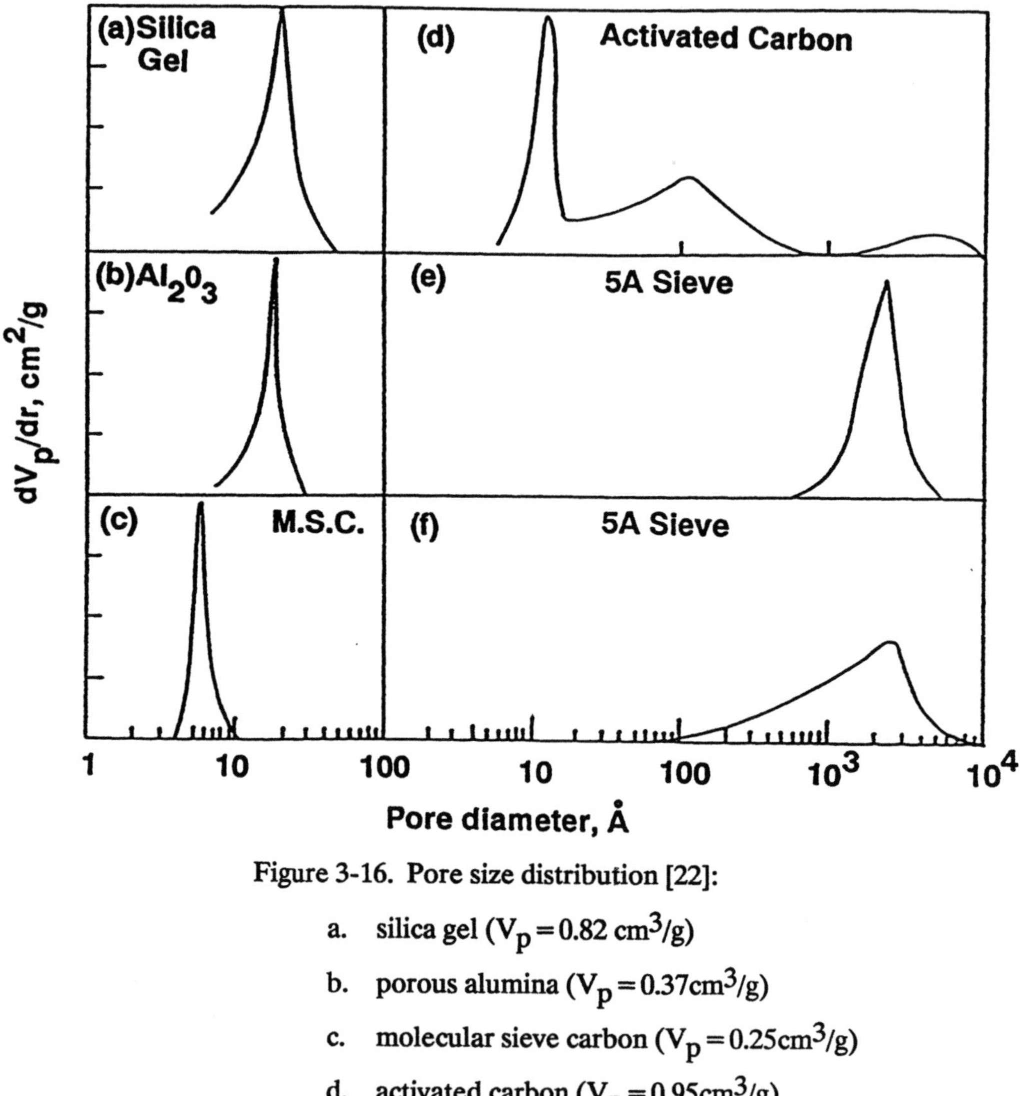

Figure 3-16. Pore size distribution [22]:

a. silica gel ($V_p = 0.82\ cm^3/g$)

b. porous alumina ($V_p = 0.37 cm^3/g$)

c. molecular sieve carbon ($V_p = 0.25 cm^3/g$)

d. activated carbon ($V_p = 0.95 cm^3/g$)

e. Davison binderless 5A molecular sieve type 625 ($V_p = 0.25 cm^3/g$)

f. Davison 5A molecular sieve type 525 ($V_p = 0.28 cm^3/g$).

$$N_b \propto b^{-(k+2)} \tag{3-32}$$

Moreover, according to Eq. 3-9, we have

$$N_b \propto b^{-d_F} = b^{-(k+2)} \tag{3-33}$$

Comparing these two expressions,

$$d_F = k+2 \tag{3-34}$$

Substituting this relationship into Eq. 3-31, we have

$$-\frac{dV}{dr^*} \propto r^{*-d_F+2} \tag{3-35}$$

In general, the mercury porosimeter data are expressed in terms of the intrusion volume, V_p, as a function of the pressure, from which the corresponding pore-size distribution is reversed. Assuming that η and ϕ to be constants in Eq. 3-24, we have

$$P \propto r^{*-1}$$

or

$$P \cdot r^* = \text{constant}$$

and therefore,

$$Pdr^* + r^*dP = 0$$

or

$$dr^* = -r^* \frac{dP}{P} \tag{3-36}$$

Suppose that the volume of pores, V, is proportional to the volume of the intrusion mercury, V_p. Then, Eqs. 3-35 and 3-36 give

$$\frac{dV}{dr^*} \propto -\frac{dV_p}{dP}\frac{P}{r^*} \propto -(r^*)^{-d_F+2} \tag{3-37}$$

and hence,

$$\frac{dV_p}{dP} \propto P^{d_F-4}, \qquad 2 < d_F \leq 3 \tag{3-38}$$

The expression implies that porosimeter method can measure the pore-size distribution, and the resultant distribution or distribution diagram in terms of dV_p/dP vs P leads to the fractal dimension of the surface of the porous material.

Application. A typical example of extracting surface fractal dimensions of char particles implementing gas adsorption methodology is given in Appendix 2 (also see [23]).

3.4 Growth Processes

Nature abounds with a countless number of shapes and forms. Yet, even today, relatively little is known about how these shapes and forms have been created. Nevertheless, we can make some elementary observations about these shapes and forms. For example, the shape of any object tends to depend strongly on the nature of the process from which it originated. Man-made objects are generally asymmetric to the eye; however, when the details are examined, it is often found that they possess essentially exact symmetries. On the other hand, naturally-formed objects frequently tend to possess apparent symmetries. Looking closer, however, we find that they are random. It is almost paradoxical that random growth processes yield objects with inherent symmetric properties [24].

In recent years, considerable interest has developed in the formation of random structures under non-equilibrium conditions; this is an important area of science which has so far received relatively little attention [25]. Computer simulation has played a key role in the development of this area; it has provided a valuable bridge between theory and experiment. We shall briefly describe two models receiving considerable attention in the field of growth processes.

3.4.1 Eden (Surface Growth) Model

This model has been proposed by Eden [26] in an attempt to simulate cell colonies. In the original version of this model, the growth process in a two dimensional plane is initiated with a single occupied lattice site, and the unoccupied surface sites (unoccupied sites with one or more occupied nearest neighbor) are occupied randomly with a probability proportional to the number of occupied nearest neighbors. In a simpler (now more common) version of the model, the unoccupied surface sites are selected randomly and occupied with an equal probability. The model is sufficiently simple so that it is possible to grow large clusters even with very crude algorithms.

Mandelbrot [5] has demonstrated that the number of occupied sites, M(r), within a distance of r from the seed site is proportional to r^{d_F}, i.e.,

$$M(r) \propto r^{d_F} \tag{3-39}$$

where d_F is the fractal dimension of the cluster. The slope of a log-log plot of M(r) versus r will yield the estimate of d_F. Figure 3-17 illustrates the interface sites for a 200,000 site cluster grown on a square lattice based on the Eden model [25]. This figure clearly indicates that the cluster is compact, i.e., its fractal dimension, d_F, is equal to its Euclidean dimension, d_E.

Since the original work of Eden, a number of modifications have been developed for various reasons. For example, if the enclosed voids (unoccupied sites surrounded by occupied sites) are not allowed to be filled in, a compact porous structure with a density of about 0.79 is formed for the case of a 2-dimensional lattice. Several other modifications of the model have been proposed recently (see, e.g., [27,28]). The most notable among them are Eden trees [29] and invading Eden [30]. All these models in which the growth probability depends on the local structure in the vicinity of a growth site yield compact structures ($d_F = d_E$).

3.4.2 Diffusion-limited Aggregation (DLA)

The importance of diffusion (Brownian motion) in colloidal aggregation has been recognized for many decades. Nevertheless, only in recent years has it been practical to carry out computer simulations of aggregation processes by resorting to Brownian (random walk) particle trajectories as first carried out by Finegold [31]. Because Finegold focused on biological applications, and performed the simulations on a rather small scale with high particle densities, reporting no quantitative results, his pioneering work has not received sufficient recognition by the physical science or engineering community.

The event contributing most to this area is the discovery by Witten and Sander [32] that a simple diffusion-limited growth model in which particles are added one at a time to a growing cluster leads to a self-similar structure having fractal properties. The model has generated considerable enthusiasm, due to the interesting scaling and universality properties associated with the structures generated by such a model [32,33], and its relevance to a wide variety of physical processes including dendritic growth [34], fluid-fluid displacement [35,36,37], colloidal aggregation [32], and dielectric breakdown [28]. The model has been investigated extensively for several years.

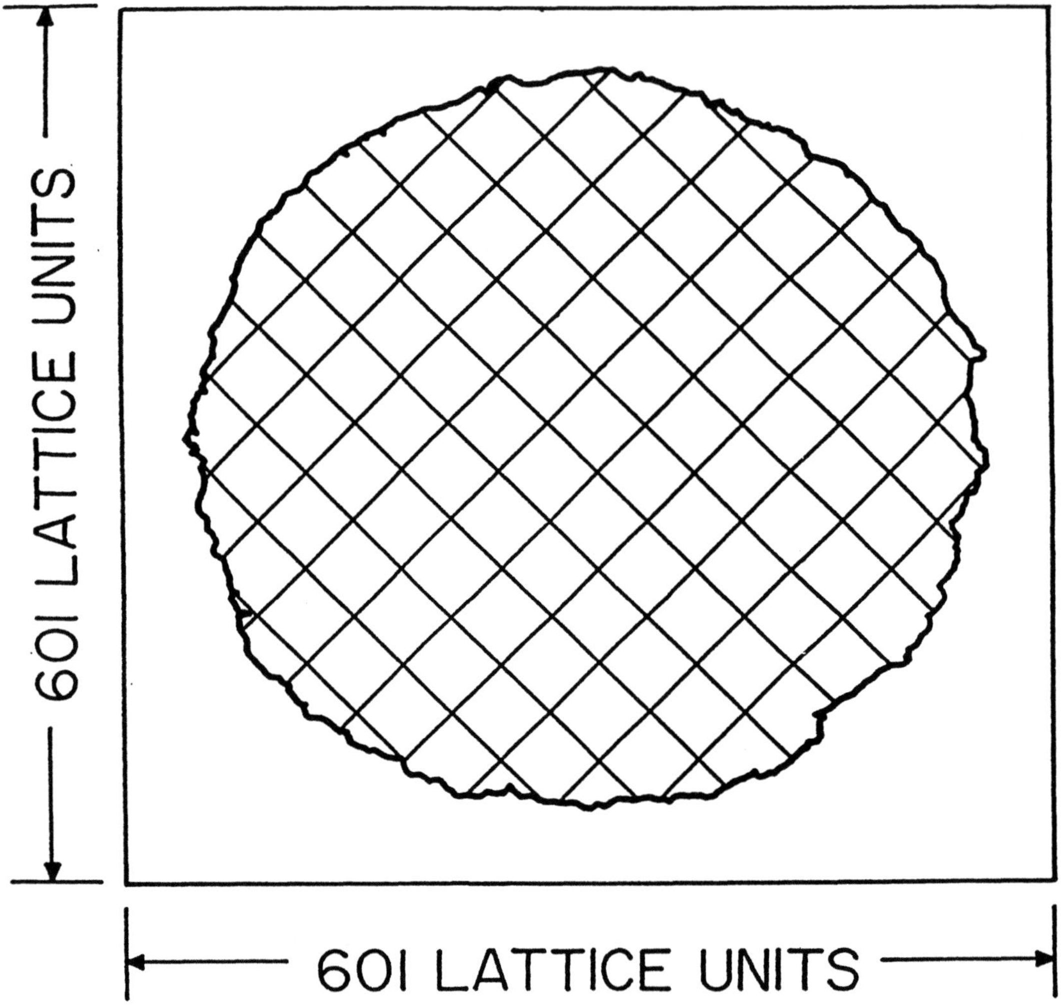

Figure 3-17. 2372 unoccupied interface sites associated with a 200,000 site cluster grown using the Eden model [25].

A simpler version of the diffusion-limited aggregation (DLA) growth process starts with a square lattice and occupies a site with a seed particle (see, e.g., [25]). A particle is then released from the perimeter of a large circle (within which the cluster is being formed) whose center coincides with the seed particle. The particle executes a random walk until it either leaves the circle or reaches a neighboring site of the seed particle. In the latter, it becomes a part of the growing cluster. This process is repeatedly performed until a sufficiently large cluster is formed. Figure 3-18 depicts such a large cluster (100,000 sites) grown on a two-dimensional square lattice based on the Witten-Sander model.

Although the growth process described above is deceptively simple, the Witten-Sander model gives rise to ramified, self-similar structures as opposed to the Eden model leading to compact structures. Witten and Sander [32] have noted that the density correlations of this model appear to have a scale-invariant form. For a fractal object, the average density-density correlation function, $<C(r)>$, has the form

$$<C(r)> \propto r^{-\alpha}, \qquad \alpha = d_E - d_F \tag{3-40}$$

over an intermediate range scale long compared to the particle or lattice size and short compared to the overall size of the cluster. To measure the density-density correlation, C(r), within an N-particle aggregates, the following definition is used:

$$C(r) = \frac{1}{N} \sum_{r'} \rho(r + r')\, \rho(r') \tag{3-41}$$

The average of C(r) over many clusters positioned at random is the average correlation function, $<C(r)>$, of Eq. 3-40. In the above equations, r is the distance separating two sites, d_E the Euclidean dimension of space, and d_F the fractal dimension of the cluster. By analyzing a variety of mass/length scaling relationships associated with these clusters, a fractal dimension of about 1.70 is obtained in two dimensions [32]. However, it can be seen from Figure 3-19 that the expected linear region with a slope of $(d_E - d_F)$ is short or absent. The upper curve in this figure shows the two points in the correlation function for the region of the cluster formed during the first 20% of the growth process. As a consequence, Eq. 3-40 does not represent the simple scaling relationship well, and a single seed from which the cluster has grown becomes a region of unusually high average density in the cluster.

The simulation of fractal-like structures with DLA models has stimulated experimental work on diffusion-limited growth processes. According to Meakin [25],

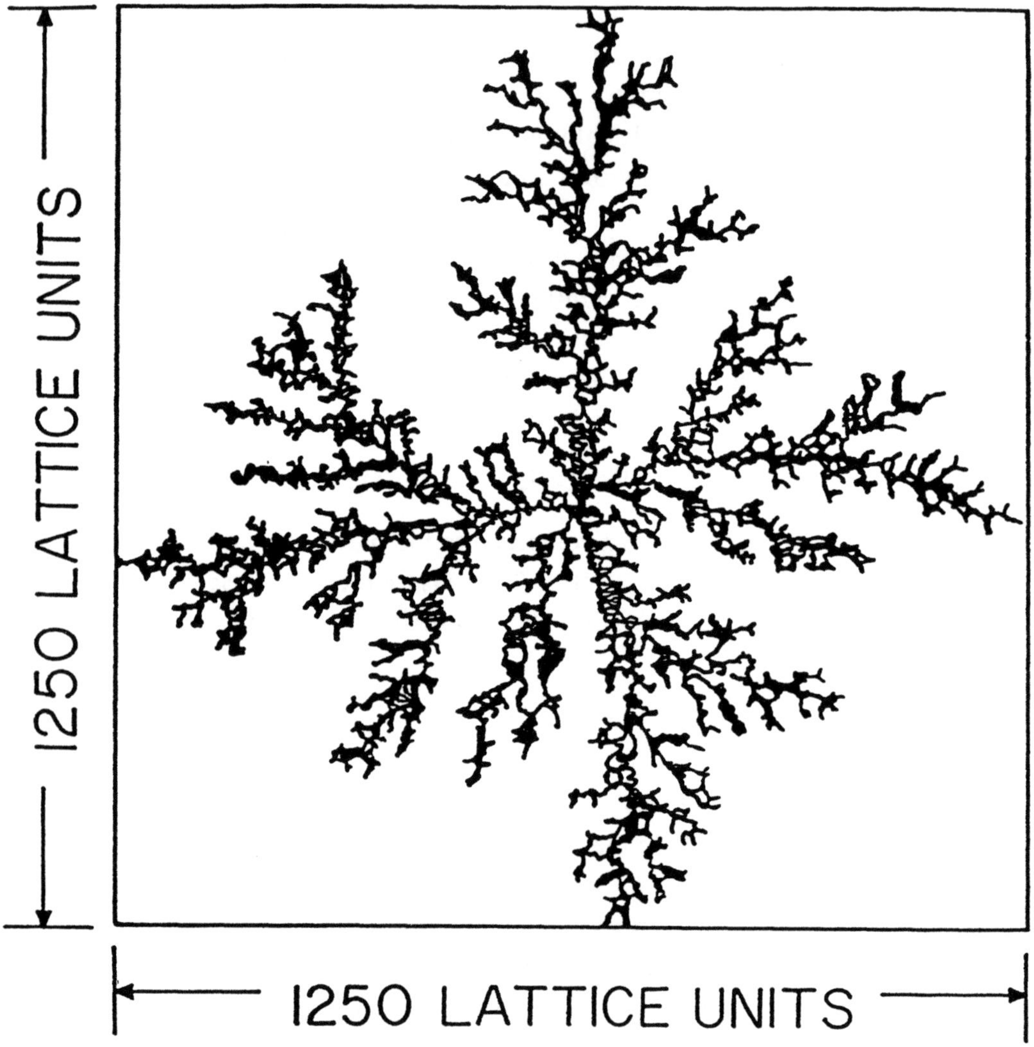

Figure 3-18. 100,000 site Witten-Sander cluster grown on a square lattice based on a semi-lattice model for DLA [25].

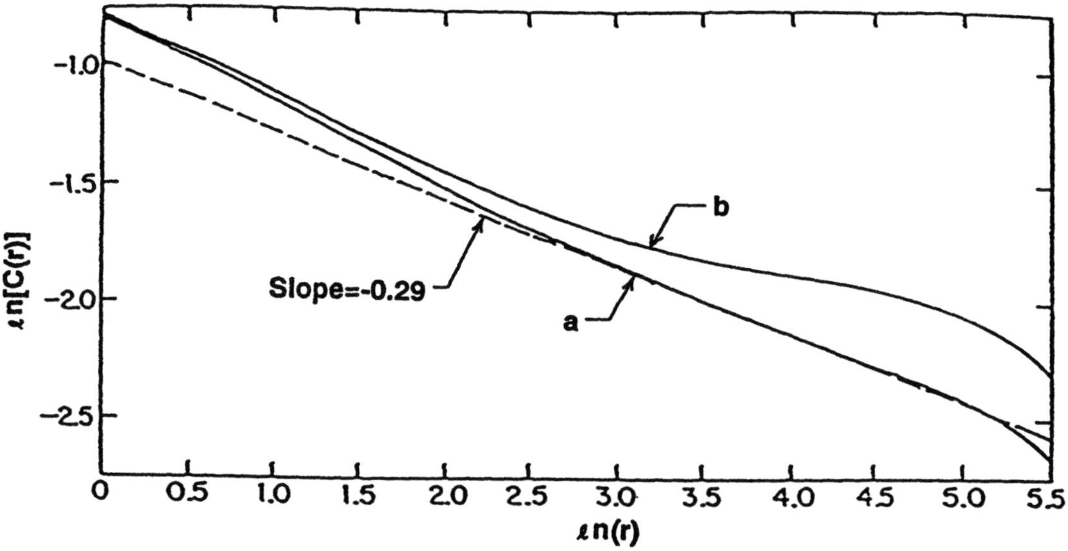

Figure 3-19. Density-density correlation functions obtained from 50,000 particle off-lattice Witten-Sander clusters [25]:

a. unrestricted
b. partially restricted.

Brady and Ball in 1984 have found a fractal dimension of 2.43 ± 0.03 for the electrode deposition of copper under diffusion-limited conditions. Similarly, Matsushita *et al.* [38] have obtained two dimensional random zinc dendrites with a fractal dimension of 1.66 ± 0.03. Moreover, structures with a fractal dimension of approximately 1.7 have been found to form under what are believed to be diffusion-limited conditions in sputter deposited niobium/germanium thin films [39]. All these experimental observations seem to be in good agreement with the computer simulated results [25].

4. FRACTALS IN TIME

This chapter contains three sections. The first outlines Mandelbrot's epoch-making work on application of the stable distribution to the modeling of change of commodity prices, the second the concept of fractional Brownian motion, and the third the concept of fractal stochastic processes leading to the notion of fractal time.

In the early 1960's, Mandelbrot published a paper related to the field of economics in which the change of commodity prices is discussed [40]. He has contended that the process of price change is discontinuous and has an infinite variance where the Gaussian distribution is inapplicable. He has also proposed a technique to analyze anomalous data based on the notion of fractional Brownian motion. Moreover, Shlesinger [41] has defined the fractal time and investigated stochastic processes, e.g., dielectric relaxation, based on this concept.

4.1 Change in Commodity Prices

The prediction of the future constitutes a central function of economics, and numerous methods have been developed for this purpose; one of them is the chart method. The so-called chartists plot the past suitably and proclaim that they can predict the future from the geometry of the resultant chart. Bachelier in 1900 counterclaimed that charting is useless because successive price changes are statistically independent (see, e.g., [40]). He has asserted that as the first approximation, any competitive price follows a one-dimensional Brownian motion, B(t), defined as

$$B(t) = Z(t+s) - Z(t) \tag{4-1}$$

based on the continuity of price changes, where Z(t) is the price at time t, and s is an arbitrary time lag; however, this Brownian motion model has failed to represent actual data well. In reality, the price changes on competitive markets need not be continuous; they are conspicuously discontinuous. Thus, a continuous process, e.g., Brownian motion, cannot account for a phenomenon characterized by extremely sharp discontinuities.

Another example, assuming the continuity of the price changes, is the method of trading to filters as depicted in Figure 4-1. In principle, a p % filter is a device that monitors a price continuously, records all the local maxima and minima, issues a buy signal when the price first overshoots the minimum by p %, and issues a sell signal when the price first reaches a local maximum minus p %. This filter assumes that price

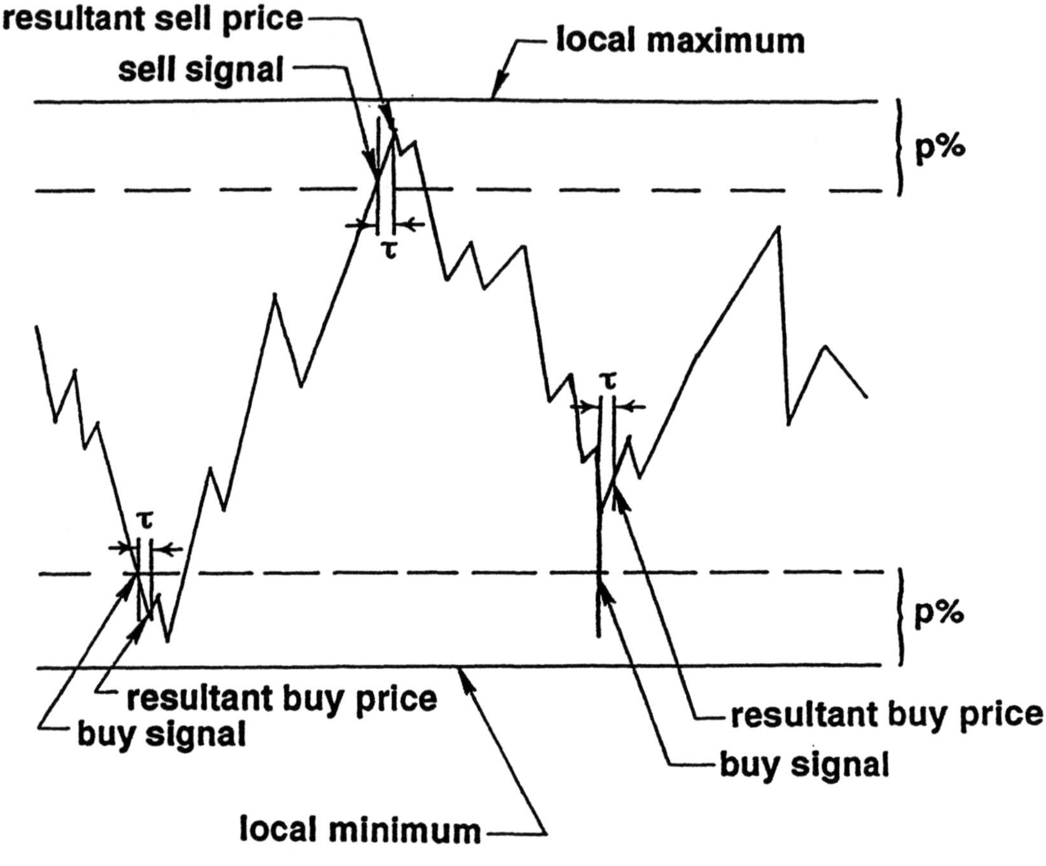

Figure 4-1. p percent filter trading [42].

changes can be treated as continuous through monitoring of the daily highs and lows. In reality the price jumps take place on many days; a buy signal will be emitted as soon as the price reaches the signal point (p % + minimum); however, the resultant buy price is often significantly higher than that assumed by Alexander [42] due to the presence of time lag between the signal point and the resultant point.

Besides the discontinuity in the price changes, the non-applicability of a Brownian model is reflected in a large variance of the price changes. This variance of sampled data is much larger than that of a Gaussian population as illustrated in Figure 4-2. These histograms demonstrate the departure from normality; two sets of data express the five and ten day price differences of monthly wool prices, 1890-1937. In each case, the continuous bell-shaped curve represents the Gaussian distributions based on the sample variance; the Gaussian distribution is much lower and flatter than the distribution of the actual data near the midpoint. The tails of the distribution of the price changes are extraordinarily long. Mandelbrot [43] has accounted for the large variance as follows:

a. Values of the variance of the price changes corresponding to different long subsamples (periods) often have different orders of magnitude.
b. As the sample size increases, the variance fails to stabilize; in fact, it tends to increase.
c. The variance tends to be influenced predominantly by a few contributing squares, i.e., the so-called outliers. When these outliers are eliminated, the estimate of dispersion often changes its order of magnitude.

This has led to the rejection of the Brownian hypothesis that the price changes are Gaussian.

The Bachelier process resulting from the one-dimensional Brownian motion (see Eq. 4-1) has been modified based on another model of the price changes, proposed by Mandelbrot [40]. In this model, Z(t) is replaced by ℓn Z(t) and B(t) by L(t,s), i.e.,

$$L(t,s) = \ell n\ Z(t+s) - \ell n\ Z(t) \qquad (4\text{-}2)$$

where s is an arbitrary time lag. Figure 4-3 represents the positive and negative tails for three series of cotton price changes. The horizontal scale, u, is marked on the lower or upper edge of the positive or negative change, respectively. The vertical scale gives the relative frequency,

$$Fr\ [\ L(t,s) > u\ ] \qquad (4\text{-}3a)$$

or

$$Fr\ [\ L(t,s) < -u\ ] \qquad (4\text{-}3b)$$

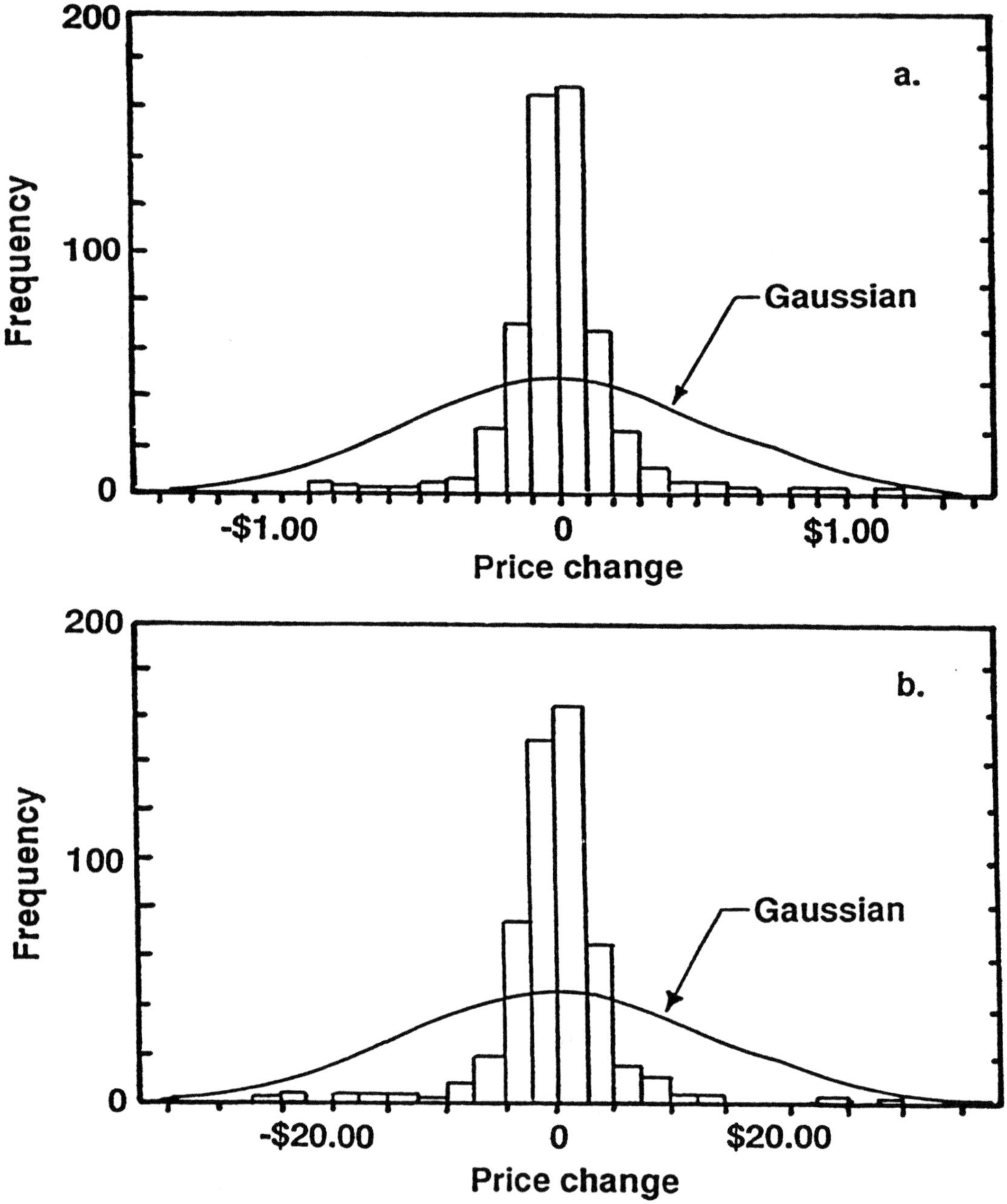

Figure 4-2. Difference between the actual data and the Gaussian distribution [40]:

a. fifth difference (s = 5 days)
b. tenth difference (s = 10 days).

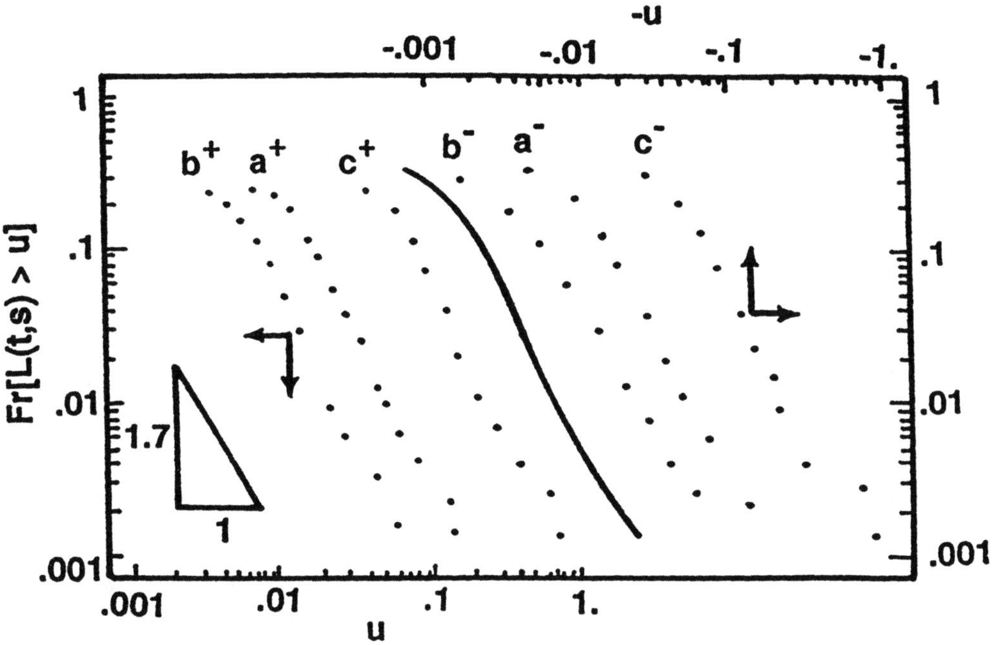

Figure 4-3. Positive and negative tails of the distribution of changes in cotton price [43]:

a. $L = \ell n\, Z(t+1\text{day}) - \ell n\, Z(t)$ where Z is the daily closing price at the New York Cotton Exchange, 1900-1905;

$$a^{+} = Fr[\, L(t,s) > u \,],\ a^{-} = Fr[\, L(t,s) < -u \,]$$

b. $L = \ell n\, Z(t+1\text{day}) - \ell n\, Z(t)$ where Z is an index of daily closing prices of cotton on various Exchanges in the U.S., 1944-1958;

$$b^{+} = Fr[\, L(t,s) > u \,],\ b^{-} = Fr[\, L(t,s) < -u \,]$$

c. $L = \ell n\, Z(t+1\text{month}) - \ell n\, Z(t)$ where Z is the closing price on the 15th of each month at the New York Cotton Exchange, 1880-1940;

$$c^{+} = Fr[\, L(t,s) > u \,],\ c^{-} = Fr[\, L(t,s) < -u \,].$$

The slope in Figure 4-3 is independent of the time lag, indicating the fractal dimension of the price changes. The population of the modified model usually has an infinite variance; it is categorized as a stable distribution (see, e.g., [43]).

4.1.1 Stable (Levy) Distribution

In 1925, Levy in his classic book "Calcul des Probabilités" first introduced the concept of stable distribution (see, e.g., [44]). Nonetheless, this distribution did not gain much attention in the fields of science and engineering because of the fact that in general the family of this distribution possesses infinite second moments and sometimes also infinite means. Only a special singular case of the family of this distribution, the Gaussian distribution, having both finite mean and variance, has became extremely popular. The plausible reasons for this are its ease of application and the fact that a substantial amount of scientific, engineering and natural data encountered in this world indeed possess Gaussian properties. It should also be noted that, with certain exceptions, stable distributions do not have explicit expressions for their densities or distribution functions.

The importance of stable distributions was reiterated by one of Levy's students, Mandelbrot [40], in connection with his work on commodity price change. The infinite mean of a stable distribution implies the absence of scale (a mean squared displacement or mean waiting time); in other words, there is no way to gauge measurements. This is consistent with stochastic processes in which self-similar clustering of a trajectory in space or events in time can result, e.g., in the Cantor set, St. Petersburg paradox to be elaborated later and Levy random walk. Mandelbrot [40] in his article on commodity price change asserted that the empirical distribution of price change is too "peaked" to be a Gaussian and that it can be better described by a stable distribution. Montroll and Shlesinger [3] have also suggested that stable distributions are often needed for describing a fractal stochastic process.

According to Levy [44], a distribution function, F(x), is said to be stable if, for any positive numbers b_1 and b_2, there exists a positive number b so that the relation

$$F\left(\frac{x}{b_1}\right) \cdot F\left(\frac{x}{b_2}\right) = F\left(\frac{x}{b}\right) \tag{4-4}$$

holds. Equation 4-4 is not so much a property of an individual distribution function F(x) but is rather a characteristic of the type to which F(x) belongs.

The defining relation, Eq. 4-4, can be expressed in terms of characteristic functions* as

$$f(b_1 t) \cdot f(b_2 t) = f(bt) \quad \text{[by convolution theorem]}^{\dagger} \qquad (4\text{-}5)$$

Since stable distributions do not always have explicit expressions for their densities or distribution functions as stated earlier, the family of these distributions are often summarized by their second characteristics¤ (or sometimes referred to as the canonical representation), $\phi(t)$, defined by

$$\phi(t) = \ell n\, f(t)$$

$$= \ell n \int_{-\infty}^{\infty} \exp(iut)\, dP_r(U,u)$$

$$= i\delta t - \gamma|t|^{\alpha}\left[1 + i\beta \frac{t}{|t|} \tan\frac{\alpha\pi}{2}\right] \qquad \text{if } \alpha \neq 1 \qquad (4\text{-}6a)$$

$$= i\delta t - \gamma|t|, \qquad \text{if } \alpha = 1 \qquad (4\text{-}6b)$$

where U is the random variable with a stable distribution. Equations 4-6a and 4-6b determine a family of distributions and their density functions $\Pr(U<u)$ and $d\Pr(U<u)$ that depend continuously upon four parameters, α, β, γ, and δ, which also happen to play the roles usually associated with the first four moments of U. First of all, α ($0 < \alpha \leq 2$) is an index of "peakedness"; if $\alpha = 1$, β must vanish. The parameter, β, is an index of "skewness" and varies from -1 to +1. If $\beta = 0$, the stable densities are symmetric. γ

---

* The characteristic function, f(t), of the distribution function F(x) is defined by

$$f(t) = E[e^{itx}] = \int_{-\infty}^{\infty} \exp(itx)\, dF(x)$$

† Convolution theorem [45]: the distribution function, F, is the convolution of two distributions F1 and F2, that is,

$$F = F_1 \cdot F_2$$

¤ The function, $\phi(t) = \ell n\, f(t)$ is called the second characteristic of the distribution function, F(x), where f(t) is the characteristic function of F(x).

expresses the scale factor, and δ indicates the location parameter. If $1 < \alpha \leq 2$ so that E[U] is finite, we have $\delta = E[U]$; if $\beta = 0$ so that the stable variable has a symmetric density function, δ is the median of U; however, δ has no obvious interpretation when $0 < \alpha \leq 1$ with $\beta \neq 0$.

As our knowledge has accumulated about the analytic properties of stable distributions, it has become clear that the complexity involved in investigating various groups of properties depends in large part on how fortuitously the parameter system is chosen. The stable distribution reduces to various well-known distributions with the right choice of parameters as illustrated below.

a) If $\alpha = 2$ and $\beta = 0$, Eq. 4-6 reduces to

$$\phi(t) = i\delta t - \gamma t^2 \tag{4-7}$$

which is the second characteristic of the Gaussian distribution with $\gamma = \sigma^2/2$ (σ^2 being the variance of the Gaussian distribution).

b) If $\alpha = 1$, Eq. 4-6 reduces to

$$\phi(t) = i\delta t - \gamma|t| \tag{4-8}$$

which is the second characteristic of a Cauchy distribution.

Now let us examine further Eq. 4-5, i.e.

$$f(b_1 t)\cdot f(b_2 t) = f(bt) \tag{4-5}$$

From the definition of f(t), we may rewrite Eq. 4-5 as

$$E(e^{itb_1x_1})\ E(e^{itb_2x_2}) = E(e^{itbx}) \tag{4-9}$$

where x_1, x_2 and x are values of the "reduced"‡ stable independent random variables, X_1, X_2 and X, respectively, i.e.,

$$X_1 \overset{d}{=} X_2 \overset{d}{=} X \overset{d}{=} U\ (\alpha,\ \beta,\ \delta = 0,\ \gamma = 1) \tag{4-10}$$

‡ If a random variable has a stable distribution such that, $\gamma = 1$ and $\delta = 0$, then the random variable is called a "reduced" variable.

Here, $\stackrel{d}{=}$ means the "same distribution." Thus, Eq. 4-9 may be expressed as

$$E [e^{it(b_1x_1+b_2x_2)}] = E [e^{itbx}]$$

or

$$b_1X_1 + b_2X_2 \stackrel{d}{=} bX \tag{4-11}$$

The characteristic functions of Eq. (4-5) may be explicitly written as

$$f(b_1t) = \exp\{-b_1^{\alpha}|t|^{\alpha} [1 + i\beta \frac{t}{|t|} \tan \frac{\alpha\pi}{2}]\} \tag{4-12a}$$

$$f(b_2t) = \exp\{-b_2^{\alpha}|t|^{\alpha} [1 + i\beta \frac{t}{|t|} \tan \frac{\alpha\pi}{2}]\} \tag{4-12b}$$

and

$$f(bt) = \exp\{-b^{\alpha}|t|^{\alpha} [1 + i\beta \frac{t}{|t|} \tan \frac{\alpha\pi}{2} \}] \tag{4-12c}$$

Thus, Eq. (4-5)

$$f(b_1t)\cdot f(b_2t) = f(bt)$$

may be rewritten as

$$\exp[-b_1^{\alpha}|t|^{\alpha} [1 + i\beta \frac{t}{|t|} \tan \frac{\alpha\pi}{2}]$$

$$\exp[-b_2^{\alpha}|t|^{\alpha} [1 + i\beta \frac{t}{|t|} \tan \frac{\alpha\pi}{2}]$$

$$= \exp[-b^{\alpha}|t|^{\alpha} [1 + i\beta \frac{t}{|t|} \tan \frac{\alpha\pi}{2}]$$

or

$$\exp[-(b_1^{\alpha}+b_2^{\alpha})|t|^{\alpha} [1 + i\beta \frac{t}{|t|} \tan \frac{\alpha\pi}{2}]$$

$$= \exp[-b^{\alpha}|t|^{\alpha} [1 + i\beta \frac{t}{|t|} \tan \frac{\alpha\pi}{2}]$$

Note that

$$b_1^{\alpha} + b_2^{\alpha} = b^{\alpha} \tag{4-13}$$

Hence, with the aid of Eq. 4-11 accompanied by its auxiliary relationship, Eq. 4-13, we may redefine a stable distribution as [45]:

The distribution of random variables X_1, X_2 and X, such that

$$X_1 \overset{d}{=} X_2 \overset{d}{=} X \neq \text{constant}$$

belongs to the family of stable distributions if and only if for any positive numbers b_1 and b_2, there exists a positive number b such that

$$b_1X_1 + b_2X_2 \overset{d}{=} bX$$

where

$$b^{\alpha} = b_1^{\alpha} + b_2^{\alpha}$$

When $\alpha = 2$, therefore, the auxiliary relation for the case of a Gaussian distribution takes the well-known form

$$b^2 = b_1^2 + b_2^2$$

4.2 Fractional Brownian Motion

One of the most useful mathematical models for analyzing time-series data is fractional Brownian motion (fBm) proposed by Mandelbrot and van Ness [46]. It is an extension of the central concept of Brownian motion that has played a significant role in both physics and mathematics. The fractional Brownian motion forms the basis for understanding anomalous diffusion and random walk which do not reconcile with the classical notion of Brownian motion. As can be seen from the sample traces of fBm in Figure 4-4 [47], the fractional Brownian motion, $B_H(t)$, is a single-valued function of one variable, t (usually time). In appearance, it is reminiscent of a mountainous horizon or the fluctuations of an economic variable or the pressure fluctuation signal from a fluidized bed. Formally, it is the increments of fBm, i.e., $[B_H(t_2)-B_H(t_1)]$, that give rise to the noise, and sum of such noises produces the traces in Figure 4-4. The scaling behavior of the different traces in the figure is characterized by a parameter H in the

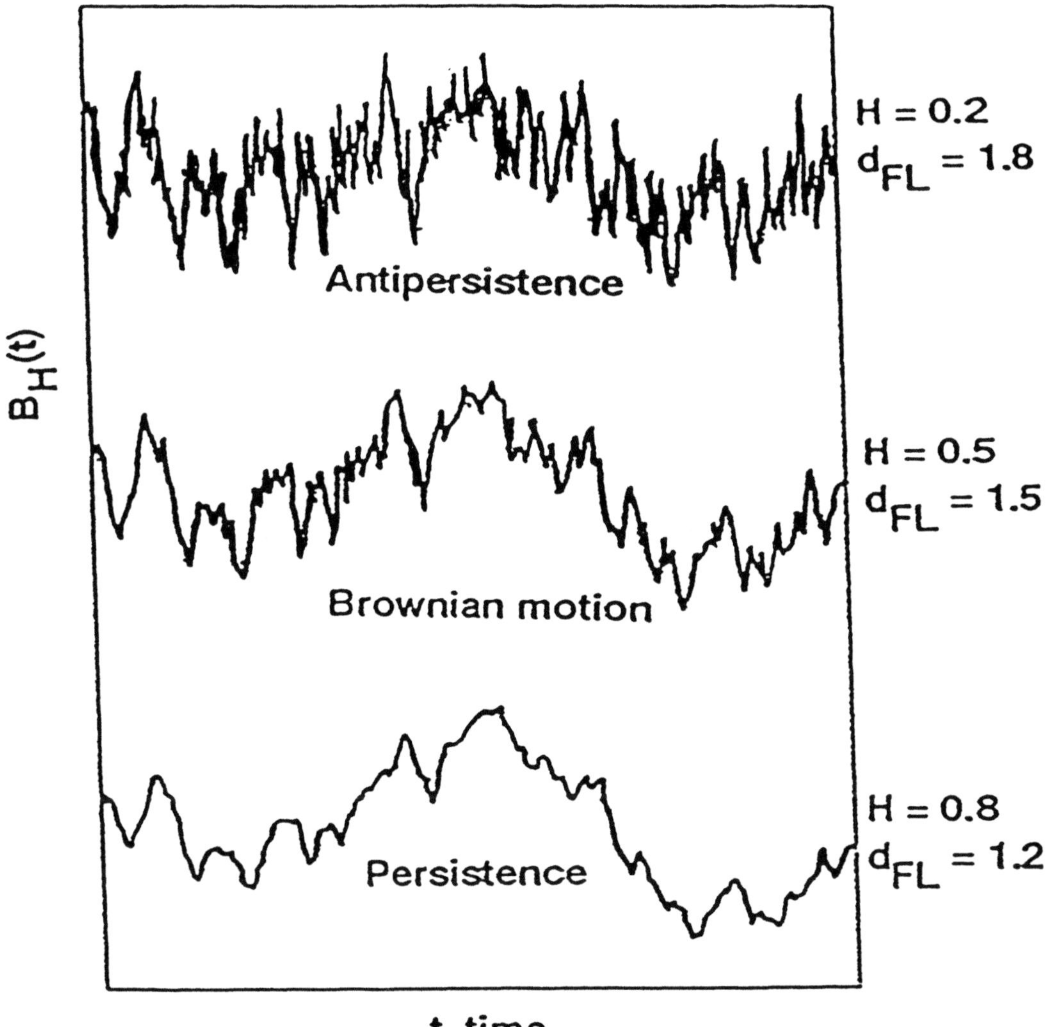

Figure 4-4. Sample plots of the fractional Brownian motion traces, $B_H(t)$, against t for different values of H and D [47]:

$$\Delta B_H \propto t^H.$$

range $0 < H < 1$. When H is close to 0, the traces are roughest while those with H close to 1 are relatively smooth. H relates the incremental change in B_H,

$$\Delta B_H = B_H(t_2) - B_H(t_1),$$

to the time difference,

$$\Delta t = t_2 - t_1,$$

through the simple scaling law

$$\Delta B_H \propto (\Delta t)^H \tag{4-14}$$

In the classical Brownian motion or random walk, the sum of independent increments or steps leads to a variation that scales as the square root of the number of steps or time increments as is well known (see, e.g., [48]); thus, $H = 1/2$ corresponds to the trace of a Brownian motion. On the other hand, the trace follows a fractional Brownian motion when $H \neq 1/2$ [46].

To differentiate between the classical Brownian motion and fractional Brownian motion, the probability distributions of the incremental jump for both types of motion are presented below [48].

For the Brownian motion, B(t),

$$P[B(t+s)-B(t)]$$

$$= \frac{1}{\sqrt{4\Pi Ds}} \exp\left[-\frac{[B(t+s)-B(t)]^2}{4Ds}\right] \tag{4-15}$$

and for the fractional Brownian motion, $B_H(t)$,

$$P[B_H(t+s)-B_H(t)]$$

$$= \frac{1}{\sqrt{4\Pi D_H s}} \exp\left[-\frac{[B_H(t+s)-B_H(t)]^2}{4D_H s}\right] \tag{4-16}$$

In the above equations the parameter, D, is the diffusivity for classical diffusion and D_H is the anomalous diffusivity for fractal diffusion; D_H is related to D by the following relationship [48,49].

$$D_H = D\, s^{2H-1} \tag{4-17}$$

Thus, for the ordinary Brownian motion,

$$\mathrm{Var}\ [B(t+s) - B(t)] = 2Ds \propto s \qquad (4\text{-}18)$$

and for the fractional Brownian motion,

$$\begin{aligned} \mathrm{Var}\ [B_H(t+s) - B_H(t)] &= 2D_H s \\ &= 2(Ds^{2H-1})s \\ &= 2Ds^{2H} \\ &\propto s^{2H} \end{aligned} \qquad (4\text{-}19)$$

In general we may say that the increment

$$[B(t+s)-B(t)]$$

of a classical Brownian motion has a Gaussian distribution with a variance proportional to lag s, whereas the increment

$$[B_H(t+s)-B_H(t)]$$

of a fBm also has a Gaussian distribution but its variance is proportional to s^{2H}; for $H = 1/2$, this obviously reduces to the result for the classical Brownian motion.

Although $B_H(t)$ is a continuous function, it is nowhere differentiable as in the case of the classical Brownian motion (see, e.g., [46]). In other words, whenever we "hit" a point of the process, it is in fact a crest or a trough, with probability of one. It is inconvenient that neither Brownian motion has a derivative. Numerous methods, many of which are not always rigorous, have been evolved to give meaning to the concept of the "derivative of Brownian motion." The resultant constructs are the so-called "Gaussian white noises." Analogous approaches can be followed with the fBm; they lead to what may be called "fractional Gaussian noises" [46].

The fractional Brownian motion, $B_H(t)$, is said to have a long-run correlation; this implies that for any three times, t_1, t, and t_2, such that $t_1 < t < t_2$,

$$B_H(t)-B_H(t_1)$$

is statistically correlated to

$$B_H(t_2)-B_H(t)$$

For $H > 1/2$, increments of $B_H(t)$ are positively correlated (persistence), i.e., an increasing trend in the past implies on the average a continued increase in the future; for $H < 1/2$, the increments are negatively correlated (anti-persistence), i.e., an increasing trend in the past implies a decreasing trend in the future. Such correlations extend to arbitrarily long time scales and have a significant effect on visual appearance of the fBm traces, as demonstrated in Figure 4-4. Note that for the singular case of $H = 1/2$, the correlation vanishes and gives rise to the well known independent increment process of the classical Brownian motion.

4.2.1 Self-affinity

At this juncture, we wish to distinguish between "self-similarity" and "self-affinity"; neither the classical nor the fractional Brownian motion is self-similar, instead, both are self-affine. Whereas the self-similar shapes repeat statistically or exactly under a magnification, the fBm traces of Figure 4-4 repeat statistically only when the t and B coordinates are magnified by different degrees. If t is magnified by a factor of r (t becomes rt), then B must be magnified by a factor of r^H (B becomes $r^H B$); for a regular random walk ($H = 1/2$), we must take four times as many steps to go twice as far. This non-uniform scaling, where the shapes are (statistically) invariant under transformations that scale different coordinates by different degrees, is known as self-affinity [50].

To reiterate, we may state that fractional Brownian motions with $H \neq 1/2$ are characterized by three properties, namely, (a) they are Gaussian processes, (b) their increments, $B_H(t+s) - B_H(t)$, constitute a stationary random process with t being a multiple of s, and (c) they are self-affine in the sense that, if time is changed in the ratio, r, and the function, $B_H(t) - B_H(0)$, is changed in the ratio, r^H, we obtain a function identical in distribution to the original $B_H(t) - B_H(0)$.

4.2.2 Discrete-time Fractional Noise

The majority of the records in time obtained in the real world are discrete time observations. $B_H(t)$ can be considered a tool for interpolating a function of discrete time (namely, the cumulative sum of Gaussian random variables) into a function of continuous time. The sequence of increments of $B_H(t)$, namely, the sequence of values of

$$\Delta B_H(t) = B_H(t+1) - B_H(t)$$

with integer values of t, is called "discrete-time fractional noise." For $H=0.5$, $\Delta B_H(t)$ reduces to a discrete-time Gaussian white noise. Thus for a record of time starting at time $t = 0$, we have (Figure 4-5),

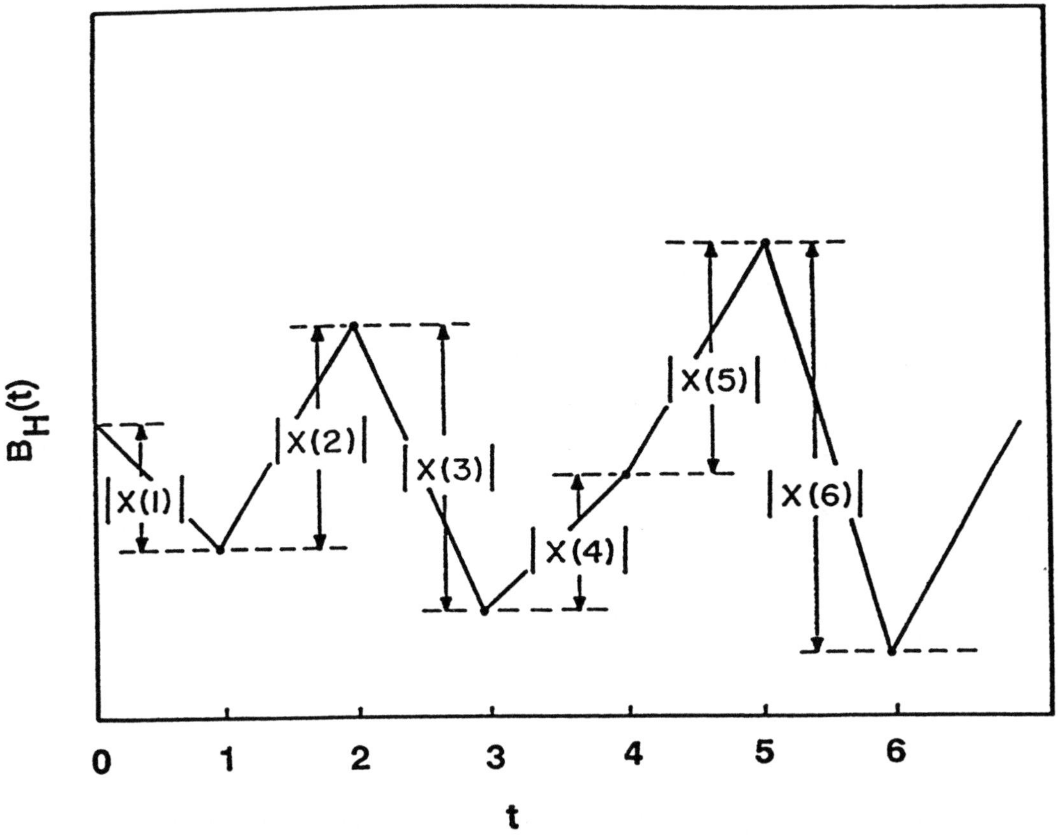

Figure 4-5. Generation of a sequence of "discrete-time fractional noise"; x(1), x(2), · · ·, from the fractional Brownian motion, $B_H(t)$.

$$X(0) = 0$$

$$X(1) = B_H(1) - B_H(0)$$

$$X(2) = B_H(2) - B_H(1)$$

$$\vdots \quad \vdots \quad \vdots$$

$$X(s) = B_H(s) - B_H(s-1)$$

Summing both sides of these expressions,

$$\sum_{u=1}^{s} X(u) = B_H(s) - B_H(0) \equiv X^*(s) \tag{4-20}$$

where s is the time lag. Thus, $X^*(s)$ is an increment of fBm and X(u) is a sequence of discrete-time fractional noise. Equations 4-19 and 4-20 imply that

$$\text{Var}\ [X^*(s)] = \text{Var}\ [B_H(s) - B_H(0)] \propto s^{2H} \tag{4-21}$$

If our recording of a signal is initiated at time t, then the definition of $X^*(s)$, Eq. 4-20, gives rise to

$$X^*(t+s) - X^*(t) = \sum_{u=1}^{t+s} X(u) - \sum_{u=1}^{t} X(u)$$

$$= \sum_{u=1}^{s} X(t+u) \tag{4-22}$$

with the variance of this quantity still remaining proportional to s^{2H} as follows:

$$\text{Var}\ [X^*(t+s) - X^*(t)] \propto s^{2H} \tag{4-23}$$

4.2.3 Rescaled Range Analysis

To obtain information about H for a given time series, we can resort to the rescaled range analysis (R/S analysis), which was originally proposed by Hurst [51]. Nevertheless, Mandelbrot and Wallis [52] were the first to apply this method to the determination of the fractal characteristics of a time series.

Let X(u) be a record containing s readings uniformly spaced from time $u = t + 1$ to time $u = t + s$; X(u) is a sequence of discrete-time fractional noise. Thus, we obtain, from Eq. 4-22,

$$\frac{1}{s}\,[X^*(t+s) - X^*(t)] = \frac{1}{s}\sum_{u=1}^{s} X(t+u) \equiv \langle X(t)\rangle_s \qquad (4\text{-}24)$$

which is the average of readings within the subrecord from time t + 1 to time t + s.

Moreover, let c(t,u) denote the cumulative departure of X(t + y) from the mean $\langle X(t)\rangle_s$, for the subrecord between time t + 1 corresponding to y = 1 and time t + u corresponding to y = u; note that by definition,

$$c(t,u) = \sum_{y=1}^{u} [X(t+y) - \langle X(t)\rangle_s]$$

$$= \sum_{y=1}^{u} X(t+y) - u\,\langle X(t)\rangle_s$$

Utilizing Eqs. 4-22 and 4-24, we obtain

$$c(t,u) = [X^*(t+u) - X^*(t)] - \frac{u}{s}\,[X^*(t+s) - X^*(t)] \qquad (4\text{-}25)$$

The sample sequential range of X(t) for lag s, R(t,s), is defined as (Figure 4-6)

$$R(t,s) = \max_{0<u\leq s} c(t,u) - \min_{0<u\leq s} c(t,u) \qquad (4\text{-}26)$$

The sample sequential variance of X(t), S^2, is defined by

$$S^2(t,s) = \frac{1}{s}\sum_{u=1}^{s} \{X(t+u) - \langle X(t)\rangle_s\}^2$$

$$= \frac{1}{s}\sum_{u=1}^{s} \{X(t+u) - \frac{1}{s}[X^*(t+s) - X^*(t)]\}^2 \qquad (4\text{-}27)$$

The ratio

$$\frac{R(t,s)}{S(t,s)}$$

is called the rescaled range [52,48].

Hurst [51] and later Mandelbrot and Wallis [52] have proposed that R(t,s)/S(t,s) is a random function with a scaling property

$$\frac{R(t,s)}{S(t,s)} \propto s^H$$

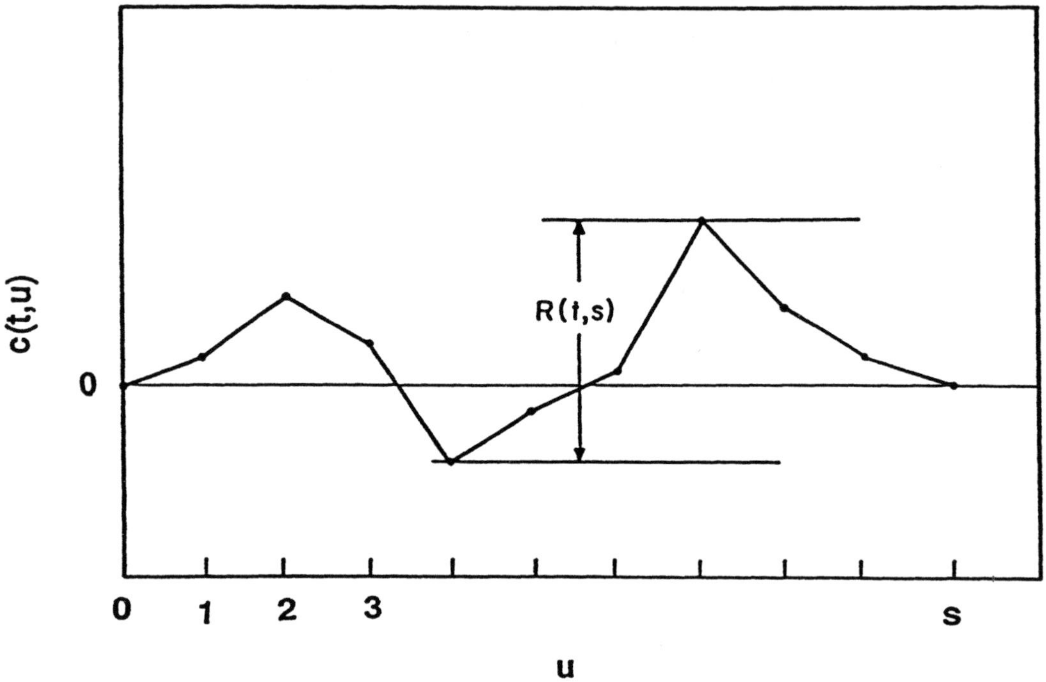

Figure 4-6. Cumulated departure from the mean, c(t,u), and the sample sequential range, R(t,s), of a time-series for the subrecord from time t + 1 to t + s:

$$c(t,u) = [x^*(t+u) - x^*(t)] - \frac{u}{s}[x^*(t+s) - x^*(t)]$$

$$R(t,s) = \max_{0<u\leq s} c(t,u) - \min_{0<u\leq s} c(t,u).$$

Feder [48] has shown that the local fractal dimension, $d_{F\ell}$, of the trace of a fBm which is a self-affine curve, is related to H by

$$d_{F\ell} = 2 - H , \qquad 0 < H < 1$$

Thus, H obtained from the R/S analysis yields the fractal dimension of the time series under investigation.

Application. Appendix 3 is a detailed example of the application of R/S analysis to characterize real time-series data (also see [53,54]). The time-series investigated in this Appendix are pressure fluctuations, produced by various multiphase flow systems.

4.3 Fractal Stochastic Processes

Mandelbrot [43] has dealt with time series, e.g., the changes in commodity prices, and fractal random walk, e.g., Levy walk. Nevertheless, the emphasis of his work appears to be on the geometry. In contrast, Shlesinger [55] has emphasized the fractals resulting from random processes and proposed the concept of fractal time.

4.3.1 Bernoulli's Scaling and the St. Petersburg Paradox

The concept of fractal time arises from the notion of Bernoulli's scaling in the St. Petersburg game. As a prelude to "playing" the St. Petersburg game [55], let us initiate ourselves to the game by learning the essence of the hypothetical "KSU" game which employs an unfair dice with different probabilities of obtaining the six numbers shown on each of its faces as listed in Table 4-1; the rules of the game are as follows:

1. The dice is thrown.
2. The player wins coins according to Table 4-1.

In this game, the expected winnings by the player, $<x>$, is

$$<x> \equiv \sum_{n=1}^{6} x_n p(x_n)$$

$$= 1\cdot\frac{1}{2} + 2\cdot\frac{1}{4} + 4\cdot\frac{1}{8} + 8\cdot\frac{1}{16} + 16\cdot\frac{1}{32} + 16\cdot\frac{1}{32}$$

$$= 3 \qquad (4\text{-}28)$$

Table 4-1. Probability and winnings in the hypothetical "KSU" game.

Number on the face of the dice n	1	2	3	4	5	6
Winnings by the player x_n	1	2	4	8	16	16
Probability $p(x_n)$	$\frac{1}{2}$	$\frac{1}{4}$	$\frac{1}{8}$	$\frac{1}{16}$	$\frac{1}{32}$	$\frac{1}{32}$

As will be demonstrated, this is a "fair" number of coins to be betted by the player. The gain by the player and that by the dealer are defined, respectively, as follows:

gain by the player = winnings - betted

and

gain by the dealer = betted - losses

For illustration, let us consider three cases in which the player bets different numbers of coins. In the first case, the number of coins betted by the player is 1 which is less than the expected winnings by the player, i.e., 3 (Table 4-2a and Figure 4-7a), and the expected gain by the player is

$$
\begin{aligned}
\langle g \rangle &= \sum_{n=1}^{6} g_n p(g_n) \\
&= (1-1)\cdot\frac{1}{2} + (2-1)\cdot\frac{1}{4} + (4-1)\cdot\frac{1}{8} + (8-1)\cdot\frac{1}{16} \\
&\quad + (16-1)\cdot\frac{1}{32} + (16-1)\cdot\frac{1}{32} \\
&= 0\cdot\frac{1}{2} + 1\cdot\frac{1}{4} + 3\cdot\frac{1}{8} + 7\cdot\frac{1}{16} + 15\cdot\frac{1}{32} + 15\cdot\frac{1}{32} \\
&= 2
\end{aligned}
$$

The expected gain by the dealer must be the negative of that by the player, i.e., -2. On the average, the number of coins betted by the player, in this case 1, is not acceptable to the dealer. In the second case, the number of coins betted by the player is 3 which is

Table 4-2. Gains of the player and the dealer in the hypothetical "KSU" game.

a. The number of coins betted by a player = 1

Number on the face of the dice n	1	2	3	4	5	6
Gain by the player $(g_n)_p$	0	1	3	7	15	15
Gain by the dealer $(g_n)_d$	0	-1	-3	-7	-15	-15
Probability $p(g_n)$	$\frac{1}{2}$	$\frac{1}{4}$	$\frac{1}{8}$	$\frac{1}{16}$	$\frac{1}{32}$	$\frac{1}{32}$

b. The number of coins betted by a player = 3

Number on the face of the dice n	1	2	3	4	5	6
Gain by the player $(g_n)_p$	-2	-1	1	5	13	13
Gain by the dealer $(g_n)_d$	2	1	-1	-5	-13	-13
Probability $p(g_n)$	$\frac{1}{2}$	$\frac{1}{4}$	$\frac{1}{8}$	$\frac{1}{16}$	$\frac{1}{32}$	$\frac{1}{32}$

c. The number of coins betted by a player = 5

Number on the face of the dice n	1	2	3	4	5	6
Gain by the player $(g_n)_p$	-4	-3	-1	3	11	11
Gain by the dealer $(g_n)_d$	4	3	1	-3	-11	-11
Probability $p(g_n)$	$\frac{1}{2}$	$\frac{1}{4}$	$\frac{1}{8}$	$\frac{1}{16}$	$\frac{1}{32}$	$\frac{1}{32}$

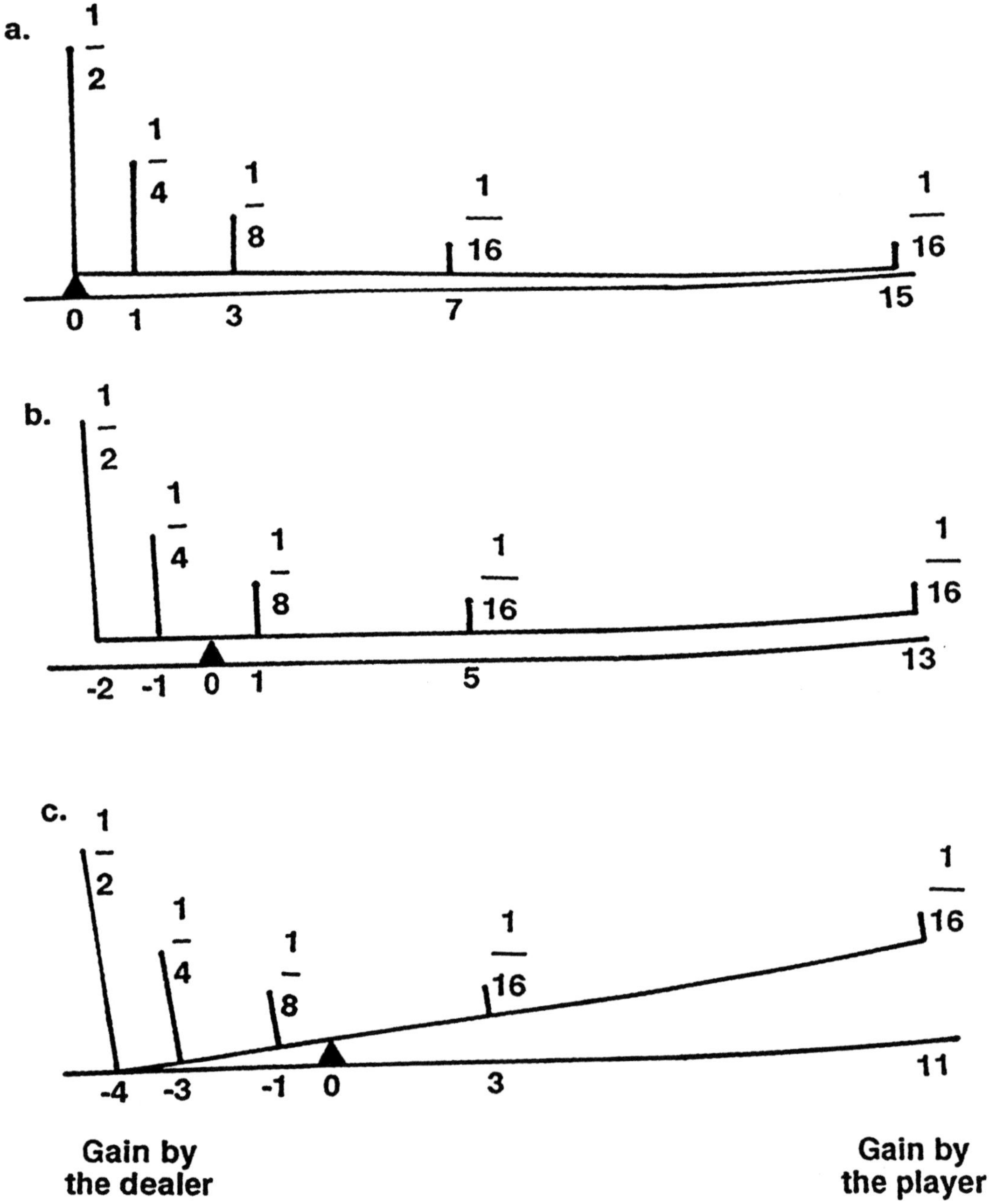

Figure 4-7. Gains of the player and dealer in the hypothetical "KSU" game:

a. number of betted coins = 1
b. number of betted coins = 3
c. number of betted coins = 5.

equal to the expected winnings by the player (Table 4-2b and Figure 4-7b), and the expected gain by the player is

$$<g> = \sum_{n=1}^{6} g_n p(g_n)$$

$$= (-2)\cdot\frac{1}{2} + (-1)\cdot\frac{1}{4} + 1\cdot\frac{1}{8} + 5\cdot\frac{1}{16} + 13\cdot\frac{1}{32} + 13\cdot\frac{1}{32}$$

$$= 0$$

Obviously, this expected gain by the dealer is also 0. Consequently, the placement of 3 by the player as the bet is acceptable to either side as predicted at the outset. In the third case, the number of coins betted by the player is 5 which is more than 3, the expected winnings by the player, (Table 4-2c and Figure 4-7c), and the expected gain by the player is

$$<g> = \sum_{n=1}^{6} g_n p(g_n)$$

$$= (-4)\cdot\frac{1}{2} + (-3)\cdot\frac{1}{4} + (-1)\cdot\frac{1}{8} + 3\cdot\frac{1}{16} + 11\cdot\frac{1}{32} + 11\cdot\frac{1}{32}$$

$$= -2$$

thereby indicating that the expected gain by the dealer is 2. Thus, the placement of 5 as the bet is not acceptable to the player but acceptable to the dealer, as expected.

In general, the player wishes to bet a smaller number of coins than 3; in contrast, the dealer wishes the player to bet a larger number of coins than 3. The ideal situation for both the player and dealer is to have identical expected gain. In the three cases illustrated, only the second case is acceptable to both. In this game, the "fair" number of coins to be betted by the player is determined to be 3, as stated earlier; this number can be regarded as the characteristic number of the game.

The St. Petersburg game is similar to the hypothetical "KSU" game; it is played as follows [55]:

1. Flip a coin.
2. If the head appears, the player wins the coin.
3. If the tail appears, flip again and again until the head appears.
4. If the head appears after $(n+1)$ flips, the player wins 2^n coins; this occurs with a probability of $(1/2)^{n+1}$, as indicated in Figure 4-8.

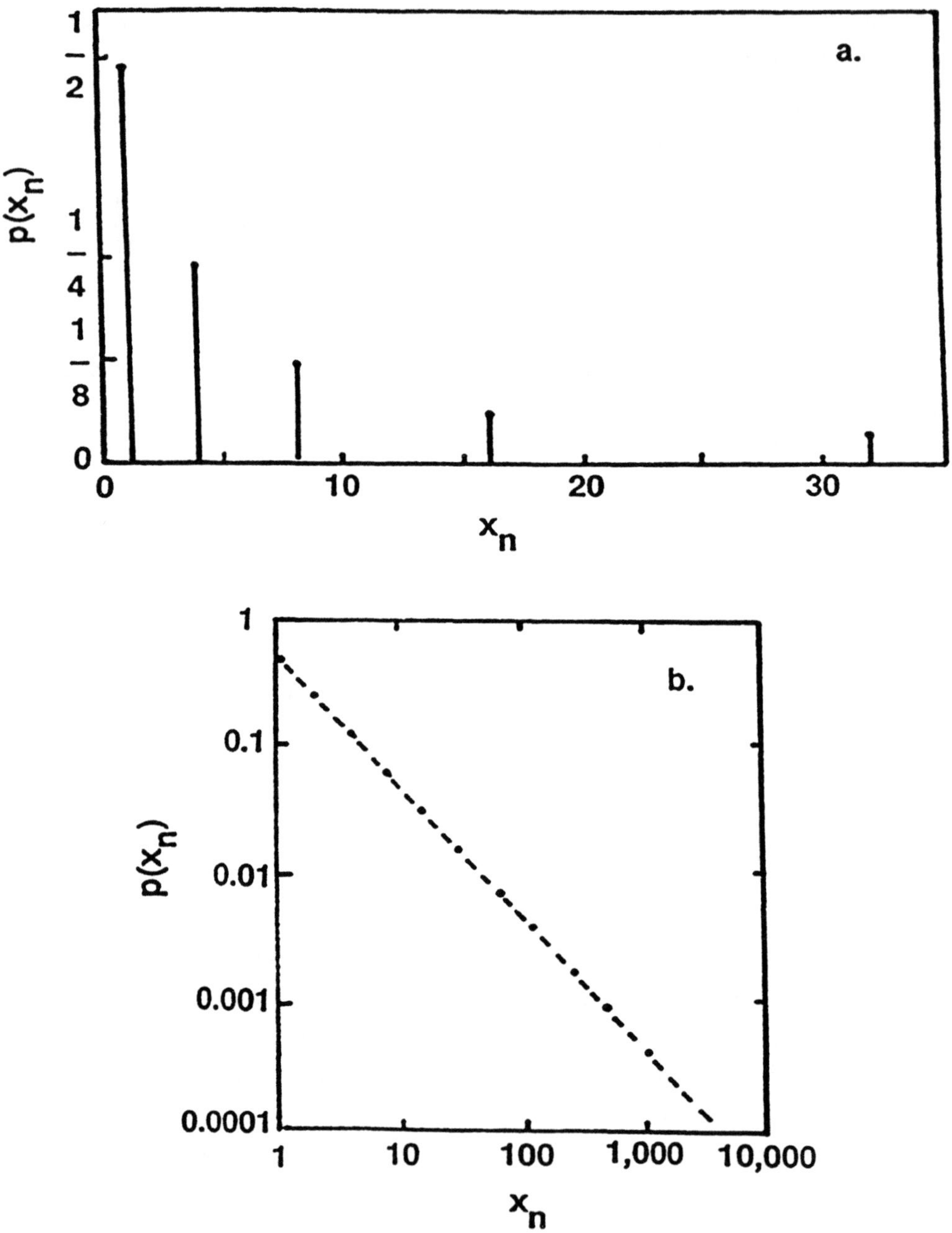

Figure 4-8. Probability density function in the St. Petersburg game [55]:

a. linear scale
b. logarithmic scale.

In this game, the mean or expected winnings by the player, $<x>$, is

$$<x> = \sum_{n=0}^{\infty} x_n p(x_n) \tag{4-29}$$

$$= 1\cdot\frac{1}{2} + 2\cdot\frac{1}{4} + 4\cdot\frac{1}{8} + \cdots + 2^n\cdot\left(\frac{1}{2}\right)^{n+1} + \cdots$$

$$= \infty$$

This is paradoxical because we cannot determine the number of coins which could be betted by the player for a "fair" St. Petersburg game. The player favors betting a smaller number of coins because the expected gain will be larger, i.e., the player can attain the maximum gain by betting the smallest number of coins. In contrast, the dealer favors the player to bet an infinite number of coins because this number is the expected winnings by the player, and thus, it is the expected loss by the dealer. Unlike the hypothetical "KSU" game there is no situation acceptable to both the dealer and player.

Unlike the distribution of the St. Petersburg game having an infinite mean value or expected gain, the distribution of the hypothetical "KSU" game has a finite mean or expected gain. Consequently, the "fair" number of coins which should be betted by the player can be determined as demonstrated earlier. The fundamental notion of the St. Petersburg game is that the player can always have the possibility of winning an order of magnitude more coins, but this occurs with an order of magnitude less in frequency or probability. This scaling idea of Bernoulli is the basis for fractal stochastic processes (see, e.g., [55]). The Cantor set illustrated in Figure 4-9 is a point set representation of the Bernoulli scaling [55]. The jump between adjacent events forms clusters or a self-similar set of waiting times. Spacings between bars become an order of magnitude smaller while their occurrences become an order of magnitude more frequent.

The second example of the scaling concept is the Weierstrass function

$$W(k) = \sum_{n=0}^{\infty} a^{-n}\cos(b^n k), \quad b > a \tag{4-30}$$

which is a continuous but nowhere differentiable function as illustrated in Figure 4-10 (see, e.g., [55]). If truncated, the sum in the right hand side of Eq. 4-30 yields a wiggly curve. Addition of a new term to this truncated sum will render the resultant curve an order of magnitude more wiggly; however, this will reduce the amplitude by an order of magnitude. The Weierstrass function is a stochastic process representation of the Bernoulli scaling; it can generate a random walk having a fractal self-similar trajectory.

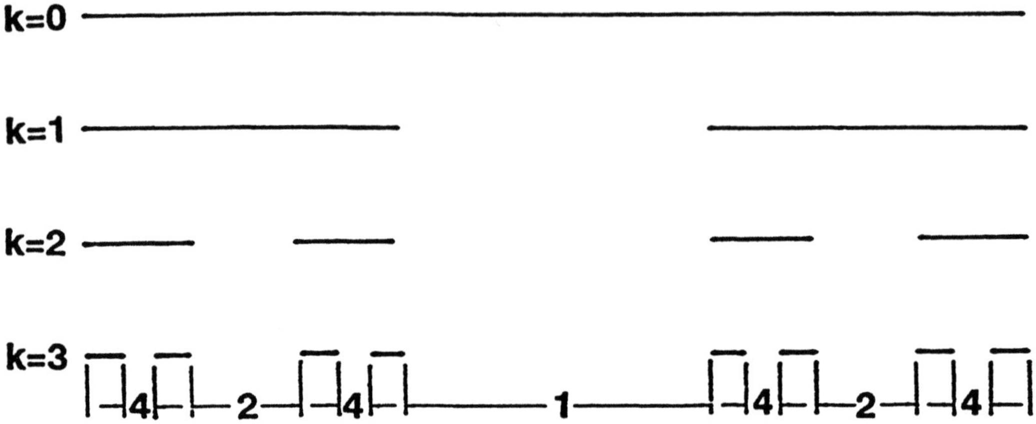

Figure 4-9. Cantor set (see, e.g., [55]).

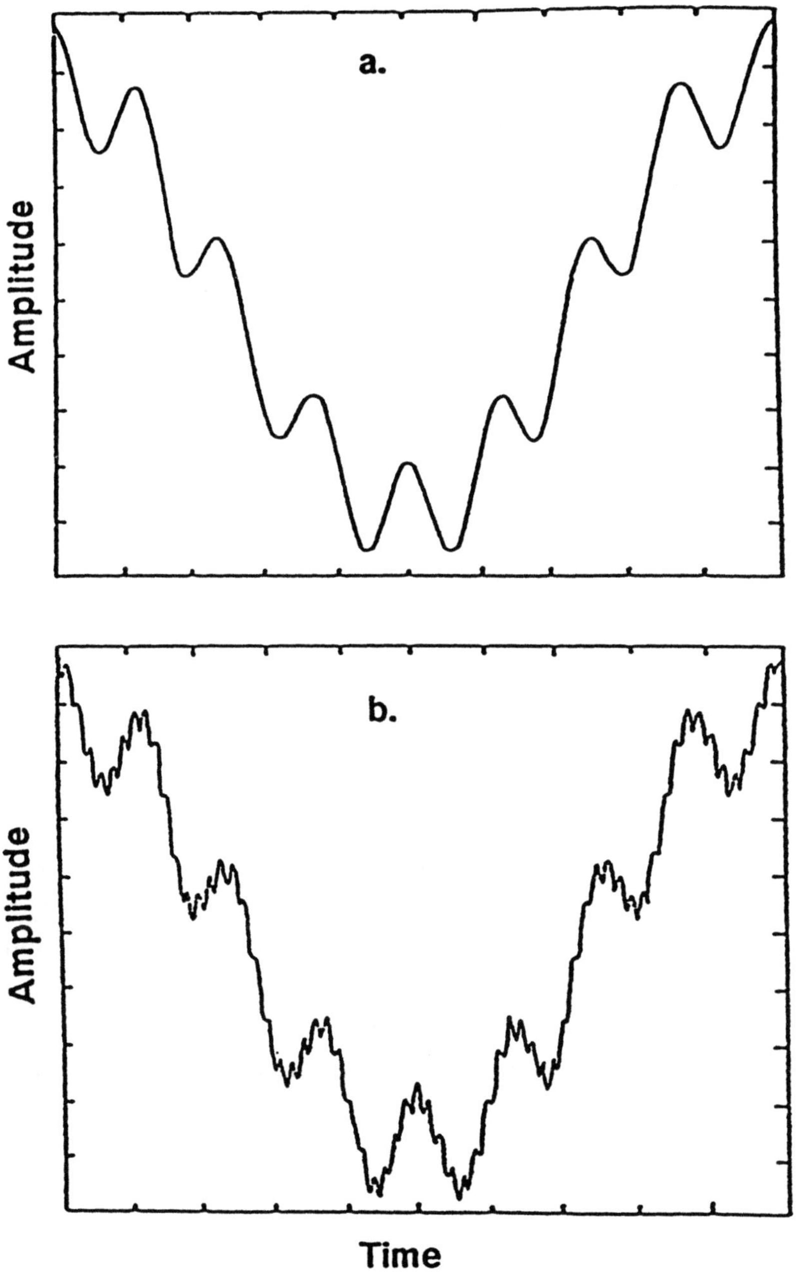

Figure 4-10. Weierstrass function (see, e.g., [55]):

a. two terms from the infinite sum
b. three terms from the infinite sum.

4.3.2 Fractal Random Walk

A random walk is represented by a sum of identically distributed random variables if the sum asymptotically approaches a limiting distribution as predicted by the Central Limit Theorem (see, e.g., [56]). This theorem states:

"Let $X_1, X_2, \cdots, X_n$ denote a sequence of random variables from a distribution that has mean μ and positive variance σ^2. Then the random variable

$$Y_n = \frac{\sum_{i=1}^{n} X_i - n\mu}{\sqrt{n}\,\sigma} = \frac{\sqrt{n}\,(\bar{X}_n - \mu)}{\sigma}$$

has a limiting distribution that is normal with mean zero and variance 1."

If the variance (the mean square displacement per jump) is finite, the Gaussian distribution of Brownian motion depicted in Figure 4-11a is obtained as

$$p(x,t) = \frac{1}{\sqrt{4\pi Dt}} \exp\left[-\frac{x^2}{4Dt}\right] \tag{4-31}$$

where p(x,t) is the probability of being at point x at time t in one-dimensional case, and D is the diffusivity (see, e.g., [57]). The Brownian motion has dominated the field of stochastic processes because of its extensive application in applied probability theory (see, e.g., [58]).

If the jump distribution has no characteristic size, i.e., when the mean square displacement is infinite, the characteristic function (Fourier transform) of p(x,t) takes the form

$$\int_{-\infty}^{\infty} p(x,t)\, \exp(izx)\, dx$$

or

$$\int_{-\infty}^{\infty} p(x,t)\, (\cos zx + i \sin zx)\, dx$$

takes the following form (see, e.g., [55]);

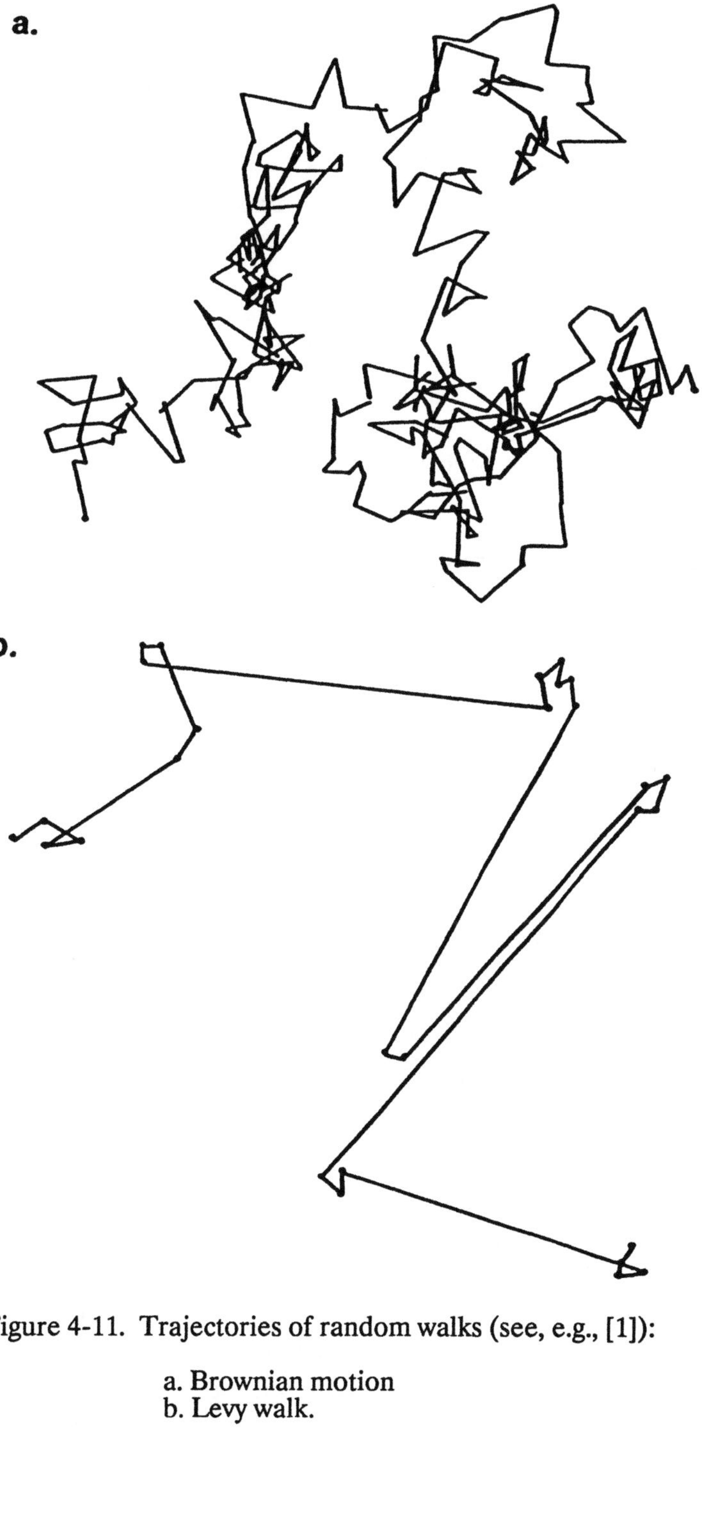

Figure 4-11. Trajectories of random walks (see, e.g., [1]):

a. Brownian motion
b. Levy walk.

$$\approx \exp(-\gamma\ |z|^{\alpha} t), \quad 0 < \alpha < 2 \tag{4-32}$$

where γ is the scale factor. Recall that Eq. 4-32 is the characteristic function of a Levy stable distribution (Eq. 4-6a). Thus, we see that discrete-time random walk with infinite variance or mean square displacement simulates a Levy or stable process.

For instance, consider a one-dimensional random walk on a lattice with a spacing of one unit between neighboring lattice points. We postulate the possible step lengths with their appropriate probabilities to be $(a>0)$

$+1$ or -1 with probability $\frac{a-1}{2a}$

$+b$ or $-b$ with probability $\frac{a-1}{2a} \cdot \frac{1}{a}$

$+b^2$ or $-b^2$ with probability $\frac{a-1}{2a} \cdot \frac{1}{a^2}$

$\vdots \qquad \vdots \qquad \vdots$

$+b^n$ or $-b^n$ with probability $\frac{a-1}{2a} \cdot \frac{1}{a^n}$

where n ranges through the positive integers. Then, the probability of displacement being x is

$$p(x) = \frac{a-1}{2a} \sum_{n=0}^{\infty} a^{-n} [\delta_{x,+b^n} + \delta_{x,-b^n}] \tag{4-33}$$

where

$$\delta_{x,+b^n} = \begin{cases} 1, & \text{if } x = +b^n \\ 0, & \text{otherwise} \end{cases}$$

and

$$\delta_{x,-b^n} = \begin{cases} 1, & \text{if } x = -b^n \\ 0, & \text{otherwise} \end{cases}$$

Intuitively, we expect the walker to make approximately "a" jumps of unit lengths (forming a cluster of sites visited) before making one of length "b". Then, approximately "a" such clusters of points visited will be generated (each separated by a distance of "b" from a neighboring cluster) before a jump of length b^2 occurs. Thus, the distribution depicted by Eq. 4-33 allows jumps of all orders of magnitude, but an order of magnitude longer jump occurs with an order of magnitude less in probability. An example of such a trajectory in a 2-dimensional lattice is depicted in Figure 4-11b.

The characteristic function of Eq. 4-33, $\tilde{p}(z)$, is

$$\tilde{p}(z) \equiv \sum_{x=-\infty}^{\infty} \exp(izx)\, p(x)$$

$$= \sum_{x=-\infty}^{\infty} \exp(izx) \frac{a-1}{2a} \sum_{n=0}^{\infty} a^{-n} [\delta_{x,+b^n} + \delta_{x,-b^n}]$$

$$= \frac{a-1}{2a} \sum_{n=0}^{\infty} a^{-n} \cos(b^n z) \tag{4-34}$$

which is precisely the Weierstrass function. It can be shown that for small z (see, e.g., [55]),

$$\tilde{p}(z) \sim \exp(-\text{const.}\,|z|^{\alpha}) \tag{4-35}$$

where

$$\alpha = \frac{\ln a}{\ln b} \tag{4-36}$$

It has been indicated that α can be regarded as a fractal dimension of the set of points visited by the random walker (see, e.g., [55]).

4.3.3 Fractal Time

The notion of fractal time has been proposed by Shlesinger [41]. Let us consider a process where the time between events (waiting time) is a random variable. Let $\psi(t)$ be the probability density function for waiting time, i.e.,

$$\psi(t)dt$$

$$= \Pr[\text{time between successive events} \in (t, t+dt)] \tag{4-37}$$

The mean, $<t>$, and the median, t_m, of the time between successive events are defined, respectively, by

$$<t> = \int_0^\infty t\,\psi(t)dt \tag{4-38}$$

and

$$\frac{1}{2} = \int_0^{t_m} \psi(t)dt \tag{4-39}$$

If $<t>$ is finite, a natural scale exists for the process; after a sufficiently long duration, the events will appear to occur at a constant rate of $<t>^{-1}$. In contrast, no natural scale exists if $<t>$ is infinite. Nevertheless, events will always occur since t_m is finite.

Suppose that three events, occurring at $t=0$, $t=\tau$, and $t=T$, result in two waiting times and that these events are treated as a renewal process homogeneous with respect to time, or time homogeneous. If T is known, the probability density function that the middle event occurs at $t=\tau$, $f(\tau)$, is

$$f(\tau) = \frac{\psi(\tau)\psi(T-\tau)}{\int_0^T \psi(s)\psi(T-s)ds} \tag{4-40}$$

where the denominator insures the proper normalization. For a purely random process with a finite mean waiting time, $1/\xi$, the waiting time distribution, $\psi(t)$, is an exponential distribution (see, e.g., [59])

$$\psi(t) = \xi \exp(-\xi t) \tag{4-41}$$

and $f(\tau)$ in Eq. 4-40 becomes a uniform distribution in the interval [0,T], as shown below

$$f(\tau) = \frac{\psi(\tau)\psi(T-\tau)}{\int_0^T \psi(s)\psi(T-s)ds}$$

$$= \frac{\xi\exp(-\xi\tau)\ \ \xi\exp[-\xi(T-\tau)]}{\int_0^T \xi\exp(-\xi s)\ \ \xi\exp[-\xi(T-s)]ds}$$

$$= \frac{\xi^2 \exp(-\xi T)}{\xi^2 \exp(-\xi T) \int_0^T ds}$$

$$= \frac{1}{T}$$

Thus, the most likely time for the middle event, i.e., the expected value of τ, is

$$\tau_{m\ell} = \frac{T}{2} \tag{4-42}$$

If a mean waiting time, $<t>$, is infinite, unlike to the above example, then T/2 is not the most likely value of $\tau_{m\ell}$; the process must be distinguishable from a pure random process whose inter-arrival time is exponentially distributed, that is, events happen randomly. The values of the most likely time for the middle event, $\tau_{m\ell}$, are far removed from T/2 but close to 0 or T. For rare events whose mean renewal rate is zero (the mean inter-arrival time is infinite), the time sequence of events must appear in self-similar clusters akin to points in the Cantor set (Figure 4-9).

Let us consider a waiting time distribution of the following form (see, e.g., [55]):

$$\psi(t) = \frac{1-p}{p} \sum_{n=1}^{\infty} p^n \lambda^n \exp(-\lambda^n t), \qquad \lambda < p < 1 \tag{4-43}$$

Note that according to this expression, the long duration between events occurs with a small probability, whereas for a pure random process any duration between events takes place with the same probability. On the average, Eq. 4-43 leads to a self-similar set of time intervals between events. The Laplace transformation of this equation results in

$$\psi^*(s) = \int_0^\infty \psi(t) \exp(-st)\, dt$$

$$= \frac{1-p}{p} \sum_{n=1}^{\infty} (p\lambda)^n \int_0^\infty \exp -(\lambda^n + s)t \, dt$$

$$= \frac{1-p}{p} \sum_{n=1}^{\infty} \frac{(p\lambda)^n}{\lambda^n + s}$$

$$= p\psi^*\left[\frac{s}{\lambda}\right] + \frac{1-p}{1+\frac{s}{\lambda}} \tag{4-44}$$

Note that

$$\psi^*\left[\frac{s}{\lambda^m}\right] = \frac{1-p}{p}\sum_{n=1}^{\infty}\frac{(p\lambda)^n}{\lambda^n + \frac{s}{\lambda^m}}$$

$$= \frac{1-p}{p}\sum_{n=1}^{\infty}\frac{(p\lambda)^{n+m}}{\lambda^{n+m}+s}\frac{\lambda^m}{(p\lambda)^m}$$

$$= \left[\frac{1}{p}\right]^m\left\{\frac{1-p}{p}\sum_{n=1}^{\infty}\frac{(p\lambda)^n}{\lambda^n+s} - \frac{1-p}{p}\sum_{j=1}^{m}\frac{(p\lambda)^j}{\lambda^j+s}\right\}$$

$$= \left[\frac{1}{p}\right]^m\left\{\psi^*(s) - \frac{1-p}{p}\sum_{j=1}^{m}\frac{(p\lambda)^j}{\lambda^j+s}\right\}$$

This scaling equation has the solution

$$\psi^*(s) = 1+s^\beta k(s) + \frac{1-p}{p}\sum_{j=1}^{\infty}\frac{(-1)^j(ps)^j}{\lambda^j-p} \tag{4-45}$$

where

$$\beta = \frac{\ell n\ p}{\ell n\ \lambda} < 1$$

Thus,

$$\langle t\rangle = -\frac{\partial\psi^*|_{s=0}}{\partial s} \longrightarrow \infty \tag{4-46}$$

and k(s) is an oscillatory function periodic in ℓn s with a period $\ell n\ \lambda$. β would appear naturally in the form of fractal dimension, and any $\psi(t)$ that behaves as $t^{-1-\beta}$ for a long time will have the same properties if $\beta < 1$. The set of intervals between events like $\psi(t)$ is similar to a cluster in the Cantor set [3].

5. FRACTALS IN CHAOS

It is now well established that seemingly innocuous dynamical systems, dissipative or not, can produce complicated phase trajectories, and eventually, chaos (see, e.g., [60]). A dynamical system, simply a system that changes its behavior over time, can become chaotic if it is sensitive to dependence on its initial conditions; this renders the system inherently unpredictable (see, e.g., [61]).

> "The equations of motion are completely deterministic, so you would think that given the initial conditions, you could predict all future behavior,"

says Harry Swinney of the University of Texas at Austin [62]; he also asserts,

> "The catch is, you have to know the initial conditions with infinite precision for a system which is chaotic."

In the physical sciences, the paragon of chaotic phenomena is turbulence (see, e.g., [61]). Thus, a rising column of smoke, the eddies behind a boat or aircraft wing provide graphic examples of chaotic motion. At low velocities the fluid patterns are quite regular and predictable. Beyond a critical velocity, however, the flow becomes turbulent. We may ask: with the predictability lost in chaotic systems, is there any order left in the system? For dissipative system, the answer is yes; there is an underlying structure to chaotic dynamics. We must search for this order in the phase space (position versus velocity), where chaotic motions exhibit a geometric property called fractal structure. Chaotic vibrations have been observed to occur in many physical systems, e.g., vibrations of buckled elastic structures, aeroelastic problems, nonlinear acoustic systems, feedback control devices, etc. [61].

Fractal dimensions play an important role in characterizing chaotic phenomena, which are usually unpredictable and nonlinear (see, e.g., [63]). This is especially true for a dissipative system where the majority of the points in its state space are transient, that is the trajectories that start there go to and stay in another part of the state space. Almost all trajectories converge towards invariant sets of relatively low dimensions with simple geometrical features termed attractors. Furthermore, the attractors are not regular surfaces in the state space but rather more complicated kinds of sets, namely, fractal sets [64].

We begin with one of the simplest attractors. Figure 5-1 depicts the attractor in a linear system, the trajectory of a pendulum. Due to friction, the amplitude decreases with time, and finally the pendulum stops. The velocity is plotted against the position of

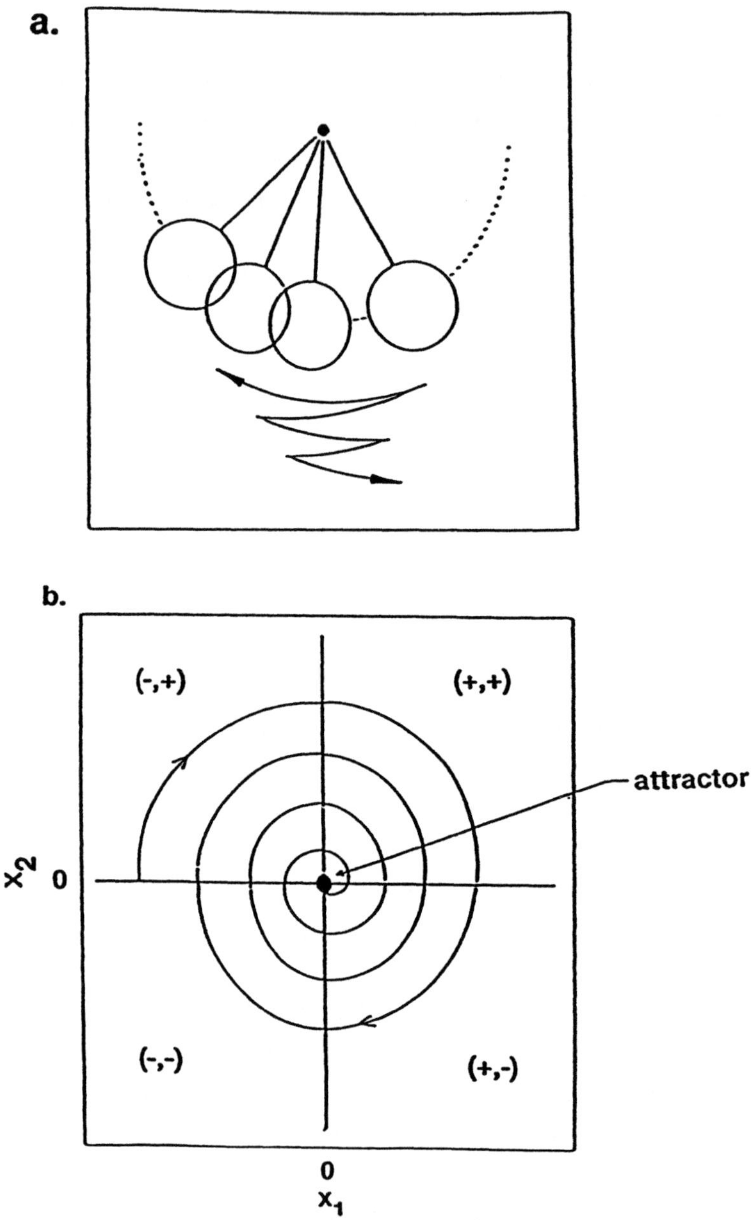

Figure 5-1. Trajectory of a pendulum (see, e.g., [65]):

a. velocity change
b. velocity vs position of a pendulum.

the pendulum. In this case, the equilibrium state or the attractor is a single point (see, e.g., [65]).

A classic example of the nonlinear system is the Van der Pol oscillator with displacement x (a vacuum tube circuit studied by Van der Pol). A mathematical model of the circuit is represented by the following equation, known as the Van der Pol equation (see, e.g., [61]);

$$\ddot{x} - \gamma\dot{x}(1-\beta x^2) + \omega x = f(t), \qquad (5\text{-}1)$$

where β,γ, and ω are parameters. The motions characterized by this expression are often called limit cycles. The phase plane trajectories of the Van der Pol equation are illustrated in Figure 5-2, in which $\dot{x}(t)$ is denoted by $x_2(t)$ and $x(t)$ by $x_1(t)$ (see, e.g., [66]). If the initial point is inside the limit cycle, the small motions spiral out to the closed asymptotic trajectory. In contrast, the motions spiral onto the limit cycle from an initial outside point. In this case, the attractor is a smooth cycle [61]. The attractor of the first example is a point, and that of the second is a smooth line; the fractal dimensions of the former and the latter are zero and one, respectively.

Another relatively simple nonlinear system is

$$x_{n+1} = 1 - ax_n^2 + y_n \qquad (5\text{-}2)$$

$$y_{n+1} = bx_n, \qquad (5\text{-}3)$$

$$n = 0, 1, 2, \cdots$$

where a and b are constants. This two-dimensional map has been proposed by a French astronomer, Henon (see, e.g., [65,67]). When a = 1.4 and b = 0.3, the plot of y against x appears as in Figure 5-3. Figures 5-4 through 5-6 exhibit the successively magnified images of the square in the preceding figure. In this case, the attractor is highly complicated and differs from those of the earlier examples. It is termed the strange attractor, and a nonlinear system having the strange attractor is termed a chaotic system. The fractal dimension of the strange attractor is often applicable for distinguishing between chaotic systems.

For a chaotic system, e.g., a turbulently mixed chemical reactor or atmosphere, the outcome of its temporal evolution for a given initial condition will be vastly different from the outcome for a slightly different initial condition after a sufficiently long time. It is nearly or totally impossible to specify at a given moment the state of such a chaotic

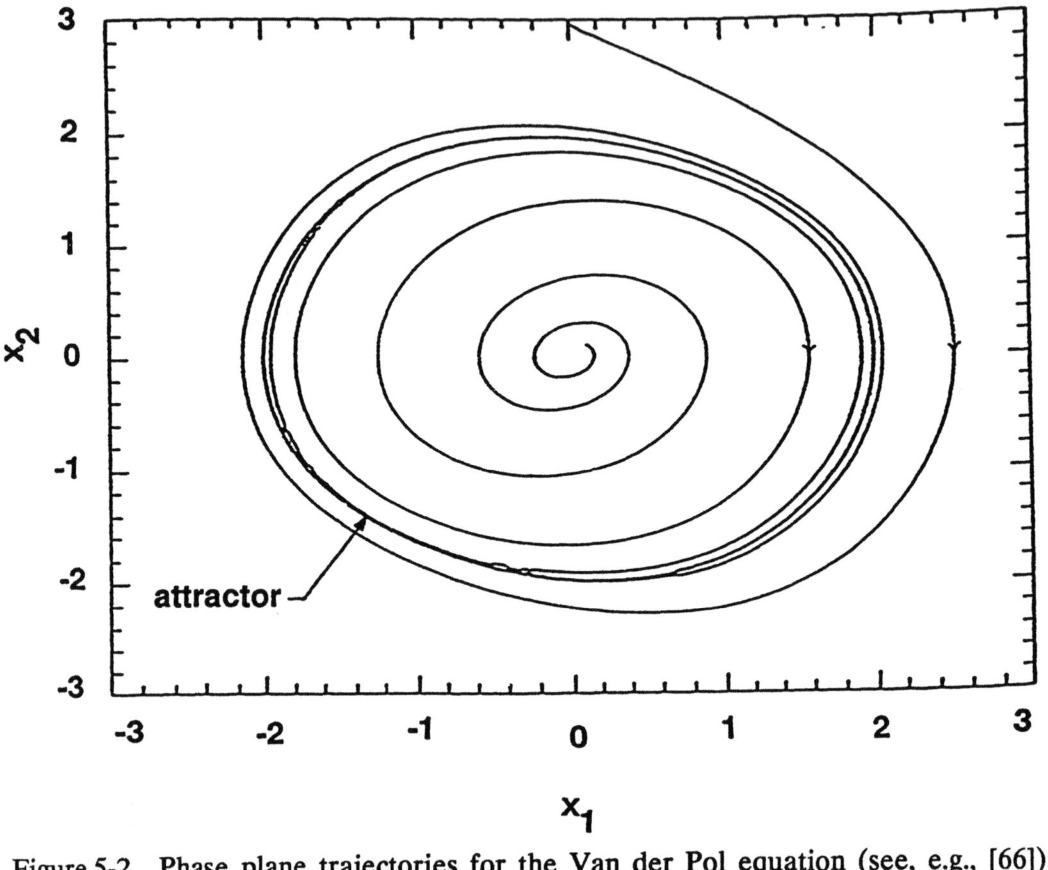

Figure 5-2. Phase plane trajectories for the Van der Pol equation (see, e.g., [66]).

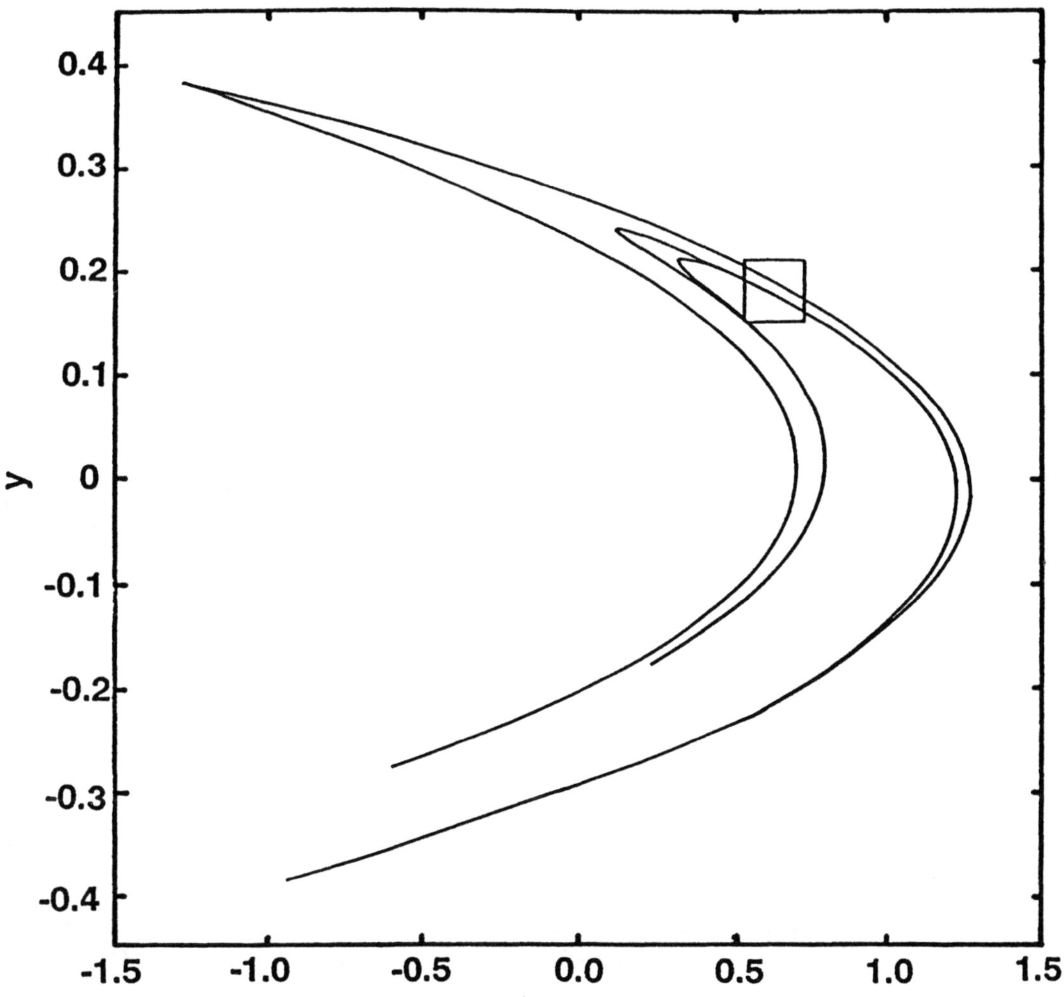

Figure 5-3. Strange attractor of the Henon map (see, e.g., [65]).

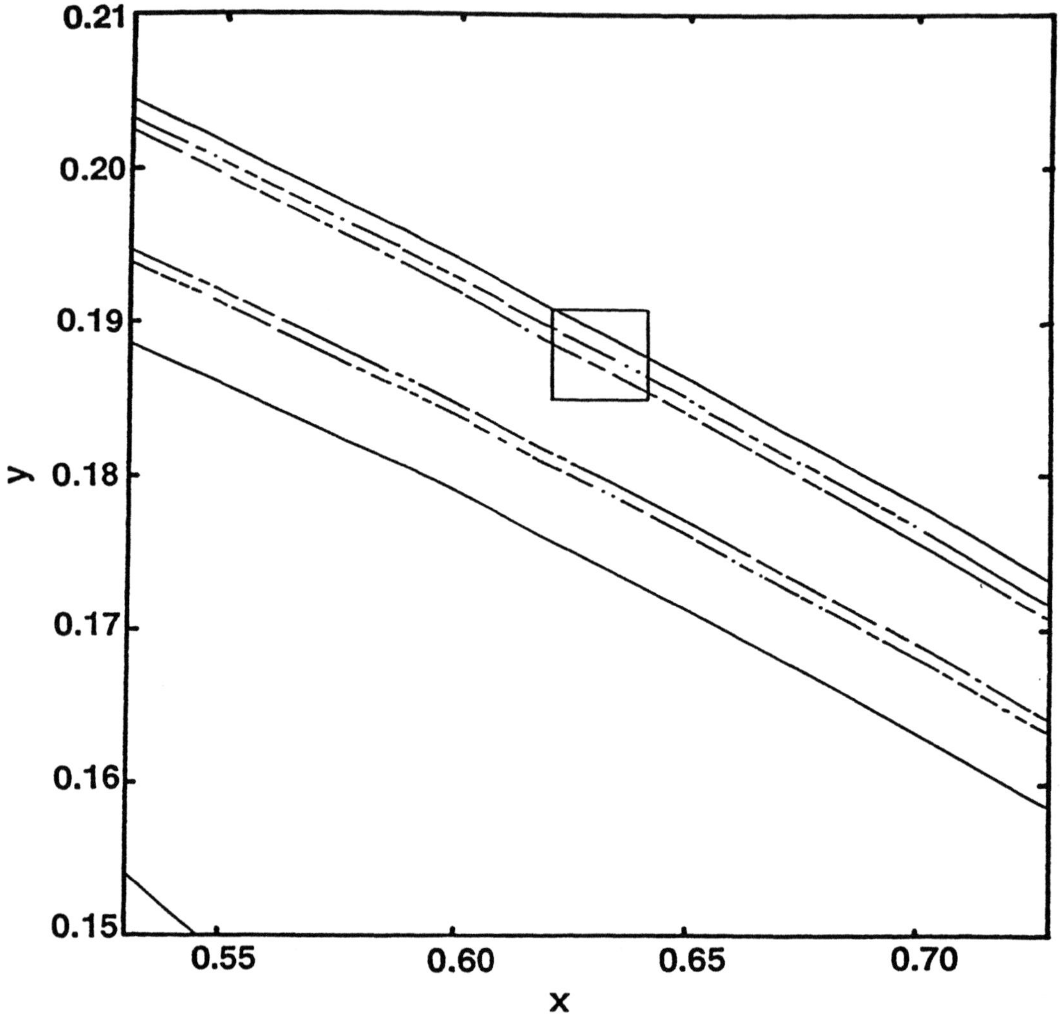

Figure 5-4. Enlargement of the square in Figure 5-3 (see, e.g., [65]).

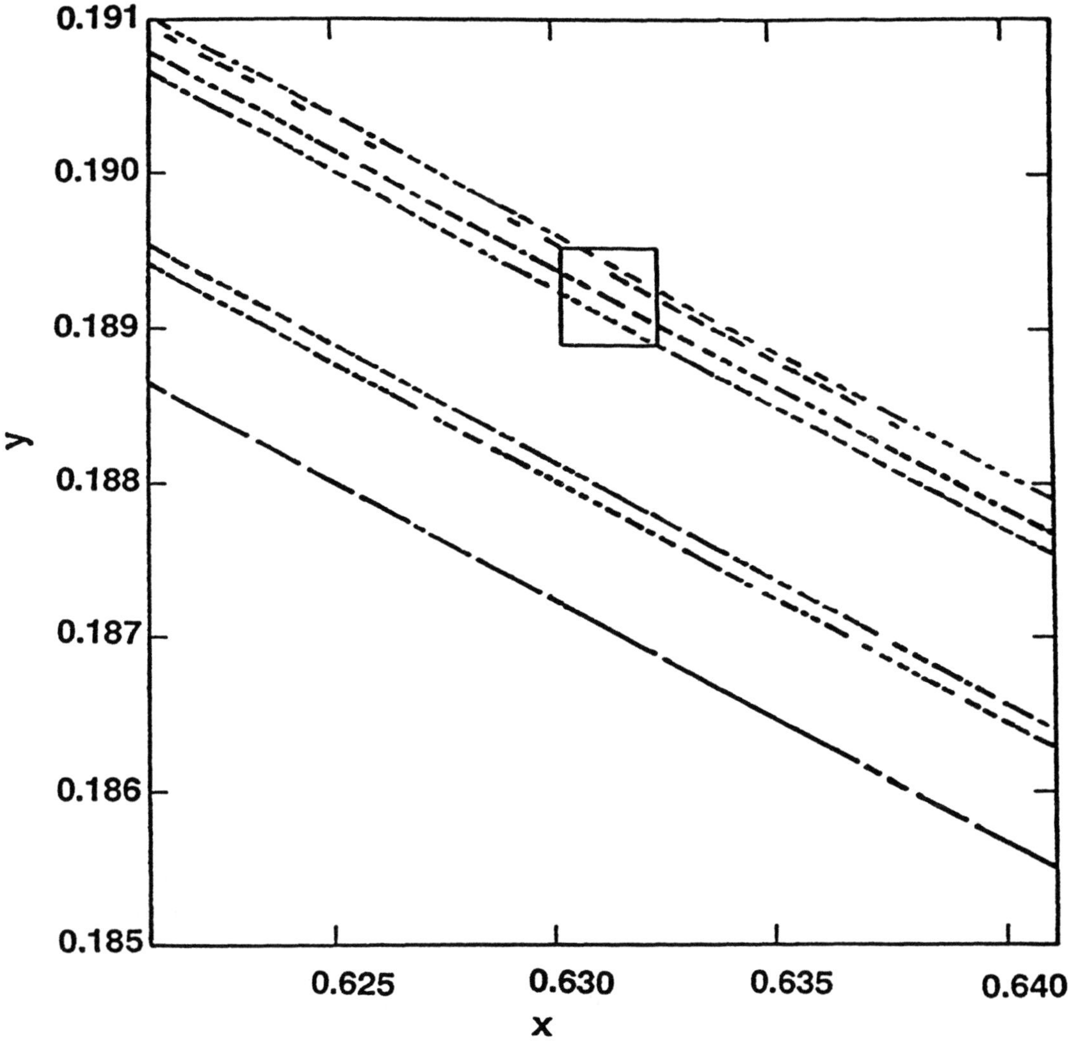

Figure 5-5. Enlargement of the square in Figure 5-4 (see, e.g., [65]).

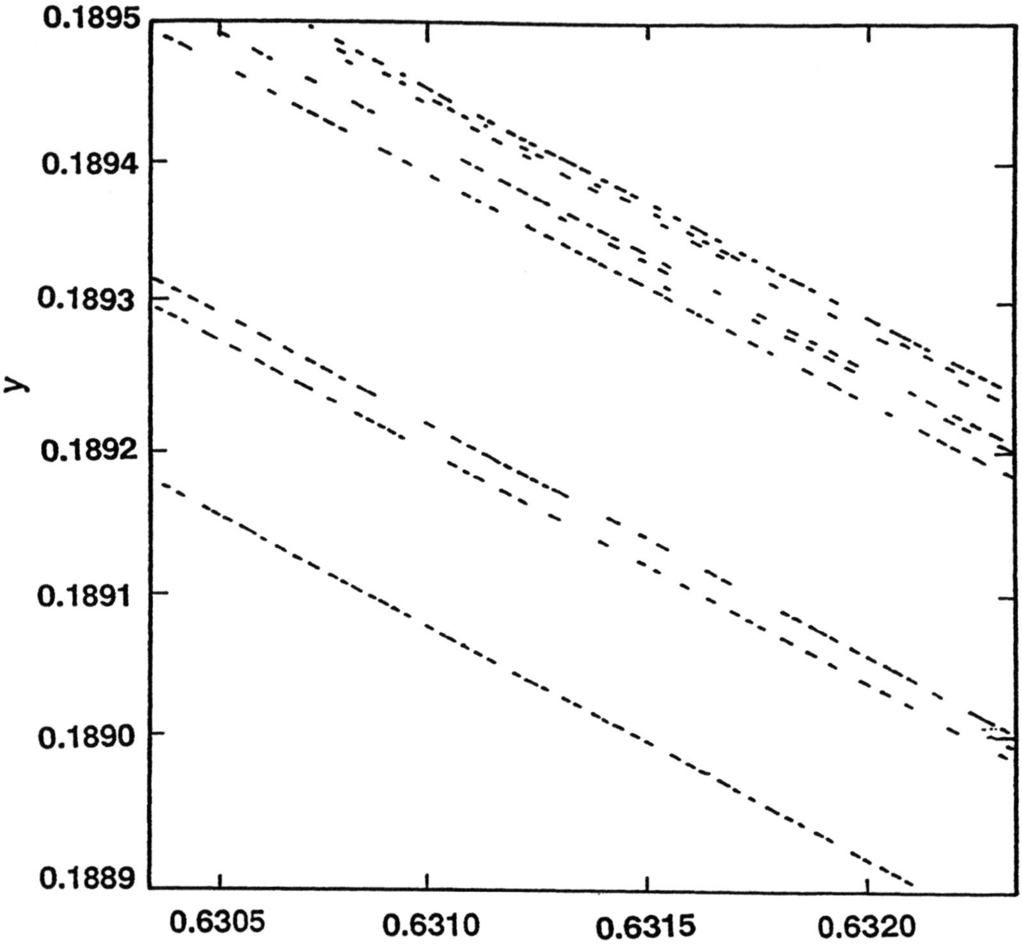

Figure 5-6. Enlargement of the square in Figure 5-5 (see, e.g., [65]).

system in terms of physical parameters, including temperature, pressure, concentration and velocity.

5.1 Quantification of Chaos

Most frequently studied physical systems, from the viewpoint of nonlinear dynamics, are forced, dissipative systems in which the low dimensional chaotic dynamics is often associated with the presence of a strange attractor in the system's phase space. Various methods provide estimates of properties of an attractor, such as the fractal dimension and Lyapunov exponents. Owing largely to the ingenuity of many experimenters the study of deterministic chaos has undergone a transition from the derivation of purely mathematical and theoretical results to the quantitative identification of chaotic features in experimental data (see, e.g., [61]).

5.1.1 Stochastic or Chaotic

Several methods are available for extracting phase-space information from experimental data; among them, the calculation of the fractal dimensions of strange attractors has probably received the most attention (see, e.g., [68,69]). The underlying notion of this approach is that a system whose dynamics is governed by a stochastic process has an infinite fractal dimension [68]; on the contrary, a finite, non-integer dimension is considered to be a clear indication of the presence of deterministic chaos. The motion of a stochastic system fills a very large-dimensional subset of its phase space, thus generating an attractor of a higher fractal dimension while that of a deterministic chaotic system asymptotically visits only a rather low dimensional subset of its phase space, thus generating an attractor of a lower fractal dimension.

Physically, the most striking contrast between a visualization based on a stochastic description and that based on deterministic chaos is the difference in the numbers of state variables characterizing the system. Suppose that a given system is dominated by low dimensional deterministic chaos; then only a relatively small number of first-order differential equations are required to express the qualitative features of its dynamics. For a nonlinear deterministic system, the number of state variables needed to describe the dynamics has been shown to be related to the attractor's dimension (see, e.g., [70]).

According to Farmer [71], the fractal dimension, d_F, of a chaotic attractor may be evaluated as

$$d_F = \lim_{\epsilon \to 0} \frac{\log n(\epsilon)}{\log (1/\epsilon)} \tag{5-4}$$

where $n(\epsilon)$ is the number of p-dimensional spheres of radius ϵ required to cover the attractor. For a chaotic system d_F is typically non-integer, finite and relatively small whereas for a stochastic system it tends to be very large. If the number of first-order ordinary differential equations necessary to simulate the dynamics of a dissipative system is N, then, the fractal dimension of the attractor would be $d_F < N$. When the system evolution is dominated by a strange attractor with a fractal dimension of d_F, then, a lower bound to the number of state variables for describing its dynamics may be fixed at d_F (the next higher integer value is used when d_F is non-integer); on the other hand, the upper limit is fixed at $2d_F + 1$. This, in turn, implies that, at most, $2d_F + 1$ ordinary differential equations (ODE) are needed to rigorously describe the system's evolution (see, e.g., [69]). Note that the relationship between the fractal dimension and the number of ODE's, required to describe the behavior of a system dominated by deterministic chaos, has a rigorous origin. In an experimental context, the existence of a low-dimensional attractor is not generally known a priori. In such a case, it is often desirable to compute the system's fractal dimension to distinguish between random noise and low-dimensional chaos.

5.1.2 Calculation of Attractor Dimension From a Time Series

A brief description is given of the method of Grassberger and Procaccia [68] for computing the correlation dimension, ν, of a strange attractor. Note that the fractal dimension, d_F, is the upper bound for ν (see, e.g., [66,68]), i.e.,

$$\nu \leq d_F \tag{5-5}$$

Given a scalar time series X(t), the first step is to reconstruct a pseudo-phase space for the system of interest. A common way to obtain a reconstructed trajectory in a p-dimensional phase space is to resort to time embedding (see, e.g., [70,72]). The experimental signal, X(t), is digitized with a time step of Δt; $(m+1)$ values of the signal, $X(i \cdot \Delta t)$, are stored for

$$i = 0, 1, 2, \cdots, m.$$

In this case, the vector time series is defined as

$$\underline{X}_i(t) = \{X(i\cdot\Delta t),\ X(i\cdot\Delta t+\tau),\cdots,X(i\cdot\Delta t+(p-1)\tau)\},\quad i = 0,1,\cdots,[m-(p-1)k] \tag{5-6}$$

where

$$\tau = k\cdot\Delta t,\ k = 1,2,\ \cdots$$

and p is the dimension of the vector, $\underline{X}_i(t)$. Thus, moving along with time t, we obtain a series of p-dimensional vectors which represent the p-dimensional portrait of the system; occasionally, p is referred to as the embedded phase-space dimension of the reconstructed attractor. Figure 5-7 illustrates a three-dimensional phase-space construction of a single time-series measurement. We can now draw a circle (for $p=2$), a sphere (for $p=3$, as shown in Figure 5-7), or a hypersphere (for $p\geq 4$) of radius r with its center at an arbitrary point in the p-dimensional phase space and count the number of points, N(r), located inside this circle, sphere, or hypersphere. This procedure is repeated for all the [m-(p-1)k] points in the p-dimensional phase space. The resulting N(r)'s are summed and then normalized to obtain what is known as the correlation integral (or the space correlation function) of the process, C(r), [68]. Hence, C(r) can be defined as

$$C(r) = \lim_{m\to\infty} \frac{1}{m^2} [\text{number of pairs } (i,j) \text{ whose distance } |\underline{X}_i(t) - \underline{X}_j(t)|<r] \tag{5-7}$$

Formally,

$$C(r) = \lim_{m\to\infty} \frac{1}{m^2} \sum_{i=1}^{m} \sum_{\substack{j=1\\ i\neq j}}^{m} H\{r-|\underline{X}_i(t)-\underline{X}_j(t)|\} \tag{5-8}$$

where $H(\cdot)$ is the Heavyside function, i.e.,

$$H\{r-|\underline{X}_i(t)-\underline{X}_j(t)|\} = \begin{cases} 1 & \text{if } r > |\underline{X}_i(t)-\underline{X}_j(t)| \\ 0 & \text{otherwise} \end{cases}$$

Grassberger and Procaccia [68] have conjectured that for small r's, C(r) scales as a power function of r, i.e.,

$$C(r) \sim r^{\nu} \tag{5-8a}$$

It follows that the correlation dimension, ν, is given by

$$\nu = \frac{\ln C(r)}{\ln(r)} \tag{5-8b}$$

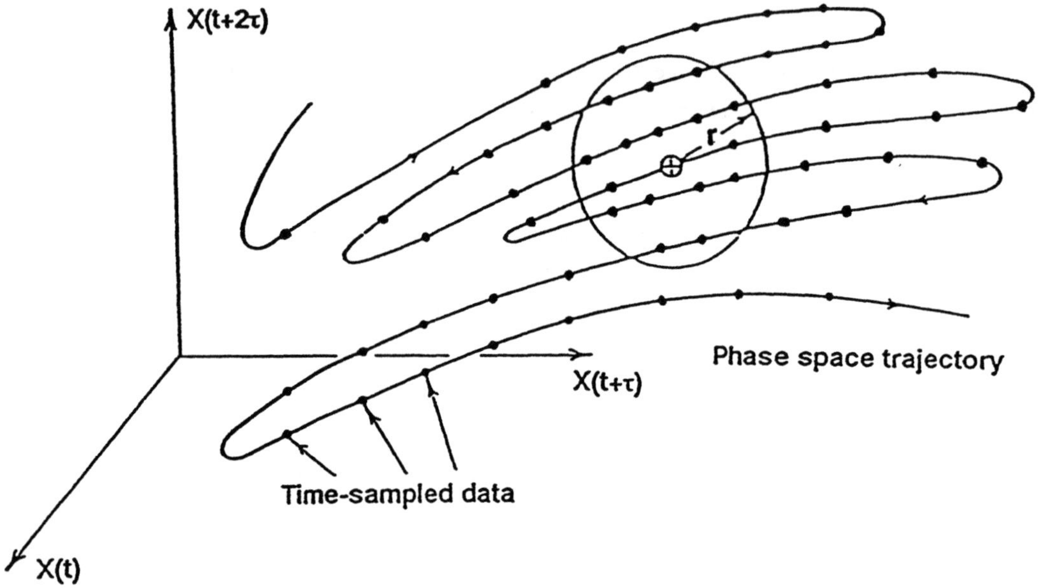

Figure 5-7. Longtime trajectory of motion in a three-dimensional phase space showing the time sampled data points and the counting sphere [66].

According to this equation a plot of $\ell nC(r)$ vs $\ell n(r)$ yields an estimate of ν for a given embedded phase-space dimension of p. When plotted against p, the correlation dimension , ν, will reach an asymptotic value if the system is governed by deterministic chaos. This asymptotic value of ν is equal to the fractal dimension, d_F, of the strange attractor [68]. In contrast, ν does not reach a finite value; if the system is governed by a stochastic process, instead it continues to increase without bound with the increase in the embedded phase-space dimension, p.

5.1.3 Lyapunov Exponent and Limits of Predictability

The Lyapunov exponent measures the speed of divergence of trajectories from nearby initial conditions. The limits of predictability are determined by this speed of divergence. According to Wolf *et al.* [73], the spectrum of Lyapunov exponents has been proven to be the most useful dynamical diagnostic for a chaotic system. A Lyapunov exponent represents the average exponential rate of divergence or convergence of nearby orbits in the phase space. Exponential orbital divergence implies that a system with a negligibly small initial difference will rapidly behave quite differently; predictive ability is rapidly lost. Any system containing at least one positive Lyapunov exponent is defined to be chaotic; the magnitude of the exponent reflects the time scale on which system dynamics becomes unpredictable.

The definition of the spectrum of Lyapunov exponents by Wolf *et al.* [73] is reiterated. "Given a continuous dynamical system in a p-dimensional phase space, we monitor the long-term evolution of an infinitesimal p-sphere of initial conditions; the sphere will become a p-ellipsoid due to the locally deforming nature of the flow. The i-th one-dimensional Lyapunov exponent is then defined in terms of the length of the ellipsoidal principal axis, $d_i(t)$, as follows:

$$\lambda_i = \lim_{t \to \infty} \frac{1}{t} \log_2 \left[\frac{d_i(t)}{d_i(0)} \right], \tag{5-9}$$

where the λ_i are ordered from the largest to the smallest." Thus, the Lyapunov exponents are related to the expanding or contracting nature of different directions in the phase space. Since the orientation of the ellipsoid changes continuously as it evolves, the directions associated with a given exponent vary in a complicated manner [73]. If $d(t_0)$ is a measure of the initial distance between two nearby starting points (Figure 5-8), the distance at a later time t_1, is

$$d(t_1) = d(t_0)\, 2^{\lambda_1 (t_1 - t_0)}, \tag{5-10}$$

where the largest Lyapunov exponent, λ_1, controls the linear extent of the ellipsoid growth given by

$$2^{\lambda_1 (t_1 - t_0)} .$$

Note that the area defined by the first two principal axes of the ellipsoid grows as

$$2^{(\lambda_1 + \lambda_2)(t_1 - t_0)} ,$$

the volume defined by the first three principal axes grows as

$$2^{(\lambda_1 + \lambda_2 + \lambda_3)(t_1 - t_0)} ,$$

and so on. It has been conjectured by Kaplan and Yorke [74] that the fractal dimension is related to the Lyapunov spectrum by the equation

$$d_F = j + \frac{\sum_{i=1}^{j} \lambda_i}{|\lambda_{j+1}|} \qquad (5\text{-}11)$$

where j is defined by the condition that

$$\sum_{i=1}^{j} \lambda_i > 0 \text{ and } \sum_{i=1}^{j+1} \lambda_i < 0$$

The calculation of dimension from this equation requires knowledge of all but the most negative Lyapunov exponents.

Wolf *et al.* [73] have presented one of the first algorithms that allow the estimation of Lyapunov exponents from an experimental time series. To measure the speed of divergence of orbits, the exponential growth rates must be averaged at a multitude of points along a trajectory as illustrated in Figure 5-8 [66]. This is initiated by specifying a point on a reference trajectory and selecting a point on a nearby trajectory. Let the initial distance between the two trajectories be $d(t_0)$. As the trajectories or orbits evolve with time, we continue to measure the distance between the two trajectories, $d(t)$, at any given time t. When $d(t)$ becomes excessively large, thus starting to depart from the exponential behavior, a nearby trajectory is identified as a replacement. Let us suppose that this occurs at the time t_1, when the distance between the two trajectories is $d(t_1)$; the ratio, $d(t_1)/d(t_0)$, is recorded. This procedure is repeated with the new or replacement trajectory and is carried on until the entire time series has been traversed. In general, let $d(t_{k-1})$ represent the initial distance between the two trajectories after (k-1) replacement steps and $d(t_k)$ the final distance between

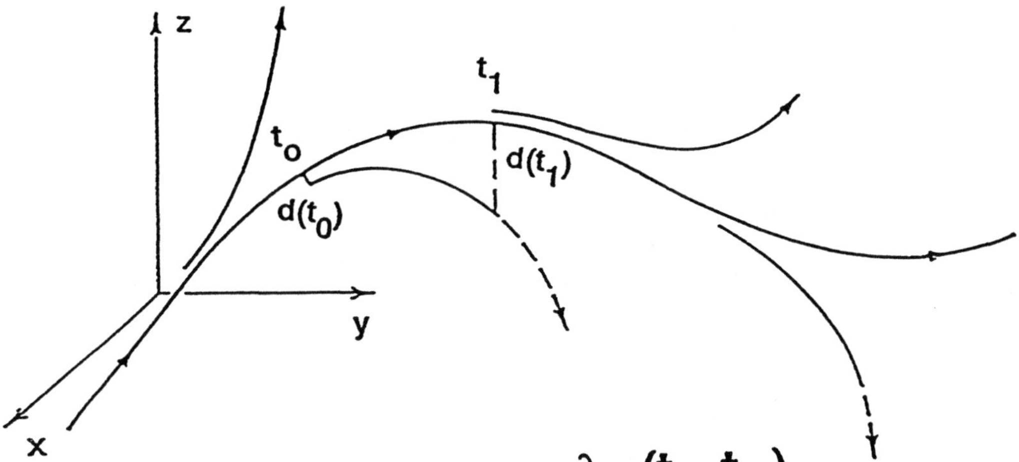

$$d(t_1) = d(t_0)2^{\lambda_1(t_1 - t_0)}$$

Figure 5-8. Change in distance between two nearby orbits for defining the largest Lyapunov exponent [66].

the two trajectories immediately before the k-th replacement step. The calculation is averaged over different regions of the phase space, and the largest Lyapunov exponent is estimated from the expression

$$\lambda_1 = \frac{1}{t_M - t_0} \sum_{k=1}^{M} \log_2 \frac{d(t_k)}{d(t_{k-1})} \qquad (5\text{-}12)$$

where M is the number of replacement steps taken. The criterion for chaos is [66]

$\lambda_1 > 0$ chaotic

$\lambda_1 \leq 0$ regular motion

The magnitudes of the Lyapunov exponents quantify an attractor's dynamics in the informational sense. The exponents measure the rate at which processes create or destroy information being expressed in terms of bits of information per unit time. Thus, if $\lambda_1 = 0.01$ bits/second, all knowledge of the initial state of the system, in terms of the bits of information, will be lost after $1/0.01 = 100$ seconds. This is the reason why the inverse of the largest Lyapunov exponent, i.e., $1/\lambda_1$, is sometimes loosely referred to as the limit of predictability of the long-term behavior of the system (see, e.g., [75]). It is worth noting that even when $\lambda_1 > 0$, we cannot ascertain if the system is chaotic or stochastic; for which we have to resort to the correlation dimensions.

6. EPILOG

Mandelbrot [1] has coined the word fractals to describe self-similar objects. He calls fractal geometry a new "geometric language." This language is geared towards the study of diverse aspects of diverse objects, either mathematical or natural, that are not smooth, but rough and fragmented to the same degree at all scales. Fractals are far more than the fantastic fruits of the cross-matching of geometric theory and computer graphics. Both the spawn and seed of a mathematical revolution, they influence and increase the range of scientific undertakings. "The mathematical concepts related to fractals have gone a long way in unifying areas of physics, chemistry, and biology that were previously obscure and couldn't be approached," says the chemist, Raoul Kopelman [76]. The concept of fractal dimension is applicable not only to spatial objects but also to random temporal processes. It can be applied to characterize stochastic time series evolving from physical, engineering or economic processes. One of the most versatile and pragmatic models of temporal fractals is the fractional Brownian motion; it encompasses the classical Brownian motion and is capable of representing long-term correlations often present in time series evolving from hydrological and other complex flow systems.

A chaotic-dynamical system is a nonlinear deterministic system highly sensitive to the initial conditions; it yields outputs not easily distinguishable from those of a stochastic process by conventional techniques. The trajectory of such a system in the phase space is characterized by the fact that it is contained in a strange attractor with a fractal dimension. It is imperative that prior to modeling, a dynamical system should be carefully analyzed to ascertain if it is governed by deterministic chaos or a stochastic process.

This monograph is an expositional review of various topics of spatial and temporal fractals, and their connections with chaos; it may be thought as a balloon ride through the wonderful and fascinating world of fractals.

A1. APPENDIX 1: Perimeter Fractal Dimension of Char Particles from a Downdraft Gasifier Through Image Analysis

Downdraft gasification in a moving bed is an extremely complex process. The gasifier performance is significantly affected by numerous parameters including the chemical and physical properties of feed materials, the geometrical forms of reacting solids, and the operating parameters. The influences of chemical and physical properties of feedstocks such as the chemical constitutes, moisture content, voidage, and bulk density and the operating parameters, such as the grate rotation speed, gas fan rotation speed, tuyere air flow, and grate open area, on the performance of a downdraft gasifier have been extensively and systematically investigated [77,78]. Nevertheless, the geometrical parameters, such as the shapes of reacting solids, have not been investigated due to the extreme difficulty in characterizing them. Some earlier comparative studies between pelletized and chipped materials have yielded interesting findings suggesting that the geometrical parameters should not be neglected (see, e.g., [79]).

In the present work, the irregular boundaries of char particles produced from a commercial-scale downdraft gasifier were characterized by resorting to the concept of fractals. Specifically, the perimeter fractal dimensions were estimated for the irregular or rugged perimeters of char particles collected under different operating conditions generated by varying the grate and gas fan rotation speeds which are the two major operating parameters influencing the performance of the downdraft gasifier [77]. The required perimeter measurements were facilitated by an image processing system.

A1.1 THEORETICAL

In this section, a brief definition of perimeter fractal dimension is outlined. Furthermore, various geometrical bases for measuring the perimeter of a rugged profile are discussed. For more details readers are referred to chapters 2 and 3 of the text.

A1.1.1 Perimeter Fractal Dimension

The perimeter of a rugged profile, P, measured as a function of the size of a probing yardstick or characteristic measurement length, λ, yields the following power law relationship if it is a fractal [5];

$$P \propto \lambda^{-d_{PF} + 1} \tag{A1-1}$$

where d_{PF} is the perimeter fractal dimension. Equation A1-1 implies that the larger the d_{PF}, the more wrinkly the object. This equation is essentially identical to Eq. 3-8 in chapter 3 of the text. Clearly, a log-log plot of Eq. A1-1, the so-called Richardson plot, yields a straight line; its slope is $-d_{PF}+1$. The selection of the characteristic measurement length or resolution, λ, is crucial for determining d_{PF} [80]. Improper selection of λ in deriving d_{PF} may generate a false prediction of the fractal dimension for the rugged profile.

According to Kaye [11], the coarsest resolution step needed to explore the rugged profile of any object, i.e., λ_{max}, should be less than 30% of the Feret diameter, F_D, defined as the maximum separation between two points along the perimeter of the rugged profile. If a circle is drawn such that it circumscribes the rugged profile, it can be seen that the largest spatial separation between two points on the rugged profile must be situated along the circumference of the circle. For convenience, we assume that

$$F_D = 2R_{max} \tag{A1-2}$$

where R_{max} is the radius of the largest circle circumscribing the rugged profile.

A1.1.2 Geometrical Bases for Measurement

If an equally-sided polygon is formed from the rugged profile with a specified value of λ, the perimeter of this profile, P, can be simply calculated as [81]

$$P = N\lambda + \ell \tag{A1-3}$$

In the above expression, N is the number of sides of the polygon and ℓ is the residual distance required to complete the polygon.

Suppose that an irregularly-sided polygon is formed from a rugged profile by linking the intersections on this profile generated when a set of parallel lines, equally spaced at a distance of λ, is laid over it. Then, the length of the i-th polygon side, denoted as ℓ_i, is the Pythagorean length between two adjacent intersections connecting the i-th side. Obviously, P is obtained as [11]

$$P = \sum_i \ell_i \tag{A1-4}$$

where i goes from 1 to N, the number of sides of the polygon.

If a rugged profile whose perimeter, P, is to be evaluated is overlaid with a rectangular grid of square tiles of dimension $\lambda x \lambda$, then P is given by [1,5]

$$P = N\lambda \quad \text{(A1-5)}$$

In this expression, N is the total number of square tiles laid along the boundary of the rugged profile.

Let R_F denote the ratio of the number of intersections between a rugged profile and a set of equally-spaced, parallel lines laid over it to the number of intersections between an equivalent smooth profile and the same set of parallel lines laid over it. Then, the greater the R_F, the more rugged the profile or the higher the d_{PF}. The equivalent smooth profile is obtained by rotating the rugged profile and identifying the maximum and minimum separations. These separations serve as the major and minor axes, respectively, for the construction of the equivalent smooth boundary. It has been determined that [82]

$$d_{PF} = 0.846 + 0.154R_F \quad \text{(A1-6)}$$

A1.2 EXPERIMENTAL

The procedures are presented for the generation and preparation of samples of char particles. A description is given for the measurement of the Feret diameter and the eventual evaluation of the perimeter fractal dimensions of these char particles.

A1.2.1 Sample Generation

The samples of char particles were obtained from the air-gasification of wood chips, or specifically black locust chips, dried to about 7% (6 ~ 8%) wet basis, in the commercial-scale downdraft gasifier depicted in Figure A1-1. The gasifier had a nominal diameter of 0.6 m. The chip bed was supported by a rotating perforated grate which was driven by a hydraulic motor. The operating range for the grate rotation speed, which could be manually adjusted with a hydraulic fluid flow splitter, was between 3 rph (revolutions per hour) and 15 rph. A hollow shaft of 7.5 cm ID supported the grate. Mounted to this shaft were a leveling rod for evening the top of the chip bed, stirrers for aiding solid flow in the bed, and tuyeres for providing air to the active zone of the gasifier.

Wood chips were fed to the top of the gasifier with a screw feeder whose speed was controlled with a hydraulic fluid flow splitter to maintain a constant chip bed level throughout the experiment. Wood chips and air flowed co-currently through several zones, i.e., the biomass-fuel, flaming-pyrolysis, char gasification, and inert char zones.

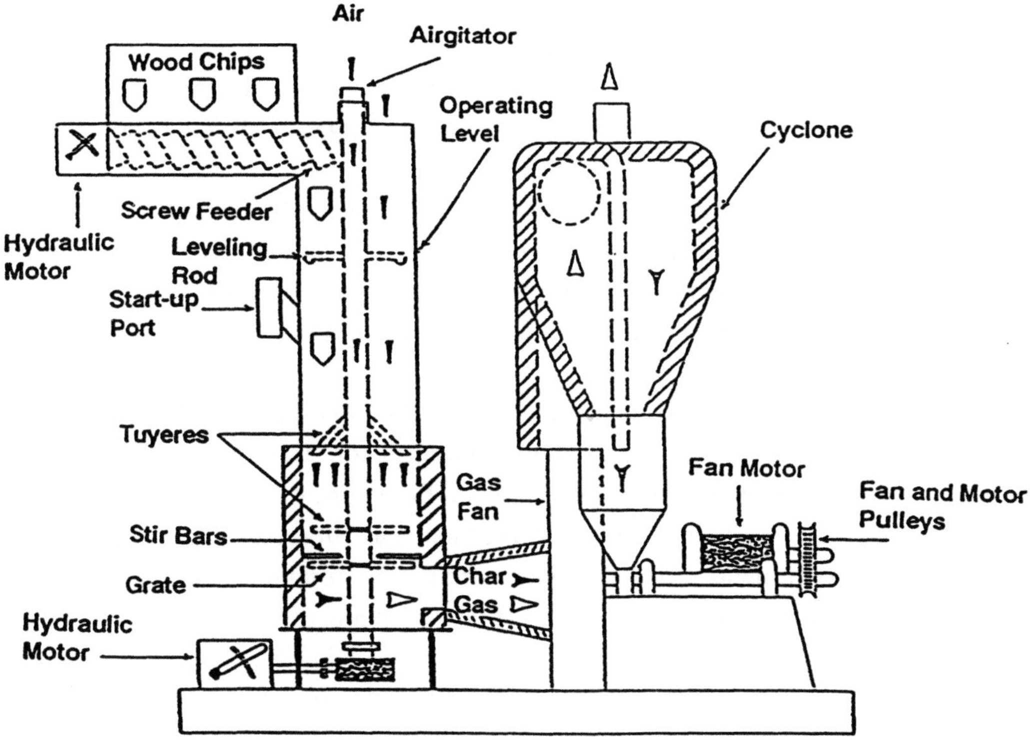

Figure A1-1. Schematic representation of the commercial-scale downdraft gasifier.

The biomass-fuel zone not only limited the air supply to the active zones; it also served as a heat insulator. Below this zone was the first active zone of the reactor, i.e., the flaming-pyrolysis zone. In this zone, pyrolysis and combustion occurred simultaneously, thus giving rise to a sharp increase in the operating temperature. Next, char particles reacted further to yield additional gases in the char gasification zone; the operating temperature began to drop slightly in this zone. The final stage was the inert char zone comprising inactive char particles and gas. Both char and gas produced were withdrawn from the bottom of this zone.

A gas fan located downstream provided suction needed to draw air into the bed through the open top and tuyeres and to discharge both char and gas effluents into the cyclone for separation. Gas fan rotation speeds ranging between 1400 rpm (revolutions per minute) and 2600 rpm could be achieved by varying the diameters of the gas fan pulley and the motor pulley.

A char sample was gathered for each of the eight operating conditions generated by changing the grate rotation and gas fan rotation speeds. Table A1-1 summarizes the operating parameters for the entire experiment. Four runs were conducted with grate rotation speeds of 3, 5, 7.5, and 12 rph and a constant gas fan rotation speed of 2200 rpm. Similarly, four experiments were performed with gas fan rotation speeds of 1400, 1800, 2200, and 2600 rpm and a fixed grate rotation speed of 3 rph. For meaningful comparison, all runs were made with the same type of wood chips, the black locust chips, whose moisture content was maintained constant between 6 and 8% wet basis as stated earlier.

Table A1-1. Summary of Gasifier Operating Conditions

Run No.	Chip Species	Moisture Content (%)	Grate Rotation (rph)	Gas Fan (rpm)
1	Black Locust	7.0	3.0	2200
2	Black Locust	7.0	5.0	2200
3	Black Locust	7.0	7.5	2200
4	Black Locust	7.0	12.0	2200
5	Black Locust	7.0	3.0	1400
6	Black Locust	7.0	3.0	1800
7	Black Locust	7.0	3.0	2200
8	Black Locust	7.0	3.0	2600

A1.2.2 Sample Preparation

The samples of char particles were collected from the clean-out port located at the bottom of the gasifier proper. Unlike the char particles exiting from the char screw conveyors, the char particles collected from the clean-out port has undergone minimum distortion such as mechanical breakage or grinding from both the gas fan and the screw conveyors. The burning char particles were rapidly quenched to eliminate further size reduction through combustion. This simple procedure can be conveniently performed to overcome the difficulty of obtaining samples directly from the active zones due to the complicated gasifier construction and high temperatures.

The major face for each char particle was photographed with a video camera and its image was digitized on a 512x512 pixel grid with a gray level of 256 using a Grinell image processing system linked to a general purpose computer (VAX). The digital image was compressed into a 256x256 pixel grid through a computer code thus reducing the computational time and storage requirement.

A1.2.3 Measurement of Feret Diameter

In a conventional procedure for determining F_D with an image analysis system, the inspected image is rotated until the maximum spatial separation between two pixels along the edge of the image is encountered; however, this procedure is often tedious and time consuming. Hence, a simple two-stage procedure was specifically established in the present work to measure the Feret diameter of all char samples generated. The procedure involves the outlining of the rugged profile and estimation of the Feret diameter, F_D.

Outlining of the rugged profile. This stage accelerates the estimation of the Feret diameter. It involves the following steps.

Step 1. Locate a dark pixel along the perimeter of the image.

Step 2. Store its coordinates and gray level in a temporary file.

Step 3. Set permanent and moving points equal to these coordinates.

Step 4. Scan the neighborhood of the dark pixel for additional dark pixels in a counterclockwise fashion.

Step 5. Select the next dark pixel; it is the one preceded by a bright pixel.

Step 6. Record its coordinates and gray level in the temporary file.

Step 7. Reset the moving pointer at these newly found coordinates.

Step 8. Compare the moving pointer with the permanent pointer. If they are equal, terminate the outlining procedure. Otherwise, return step 4.

Estimation of the Feret diameter. In this stage, all coordinates stored in the temporary file from the outlining stage are made available for the estimation of F_D. It is performed according to the following steps.

Step 1. Construct a rectangle circumscribing the outlined profile.

Step 2. Locate the center of the rectangle.

Step 3. Calculate the distance between each pixel on the boundary and the center.

Step 4. Compare all distances to search for the maximum distance, R_{max}.

Step 5. Estimate the Feret diameter, F_D, from R_{max} by resorting to Eq. A1-2.

Figure A1-2 illustrates the difference between the actual and estimated F_D's for a rugged profile. The estimated F_D tends to slightly over-estimate the actual F_D. As stated previously, F_D provides the upper bound of the characteristic measurement length, λ, for deriving the perimeter fractal dimension through Eq. A1-1. Because of the approximate nature of Eq. A1-2, we have conservatively estimated the coarsest characteristic measurement length, λ_{max}, as

$$\lambda_{max} = 0.25F_D \tag{A1-7}$$

The value of λ_{max} derived from this relation tends to be below the value estimated from the 30% rule [11].

A1.2.4 Evaluation of Perimeter Fractal Dimension

The perimeter fractal dimensions of char particles were determined with an empirical but efficient method. It was developed by slightly modifying the method leading to Eq. A1-6. In the present method, instead of obtaining the equivalent smooth profile for the rugged profile under investigation, two circles, the largest one circumscribing and the smallest one inscribing the rugged profile, respectively, were determined for calculating an average value of R_F, defined previously when deriving Eq. A1-6. Obviously, both the largest and smallest circles can be recovered through the Feret diameter estimation procedure. Note that the largest circle has been adopted to evaluate the Feret diameter described earlier. During the process of distance comparison as outlined in step 4 of the estimation of Feret diameter, the smallest circle is obtained from the minimum distance between the boundary and the center of the rectangle. For

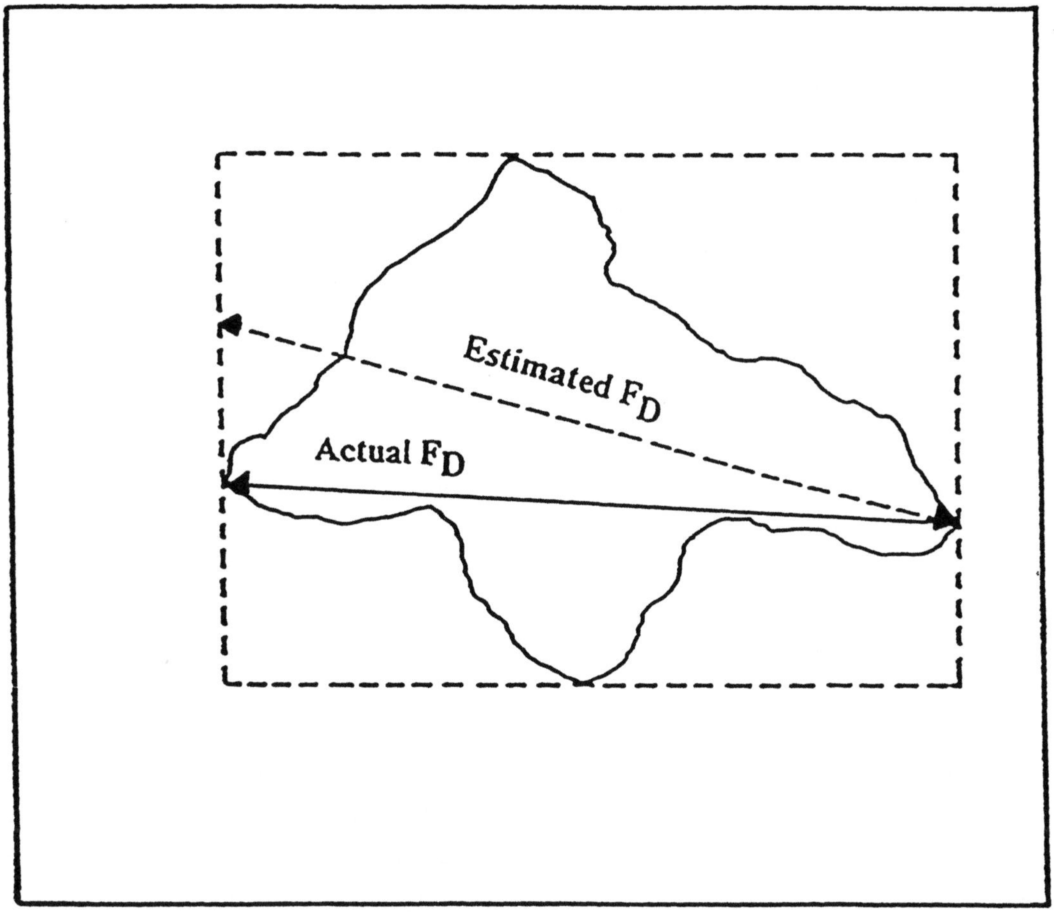

Figure A1-2. Difference between the actual and estimated F_D's for a rugged profile.

each circle, the ratio of the number of intersections between the set of equally-spaced, parallel lines and the rugged profile to the number of intersections between the similar set of lines and the circle was evaluated. A geometrical mean value, R_F, is evaluated from the ratio of the largest circle and that of the smallest circle.

To derive an empirical equation relating d_{PF} to R_F, which is analogous to Eq. A1-6, data were generated with the present image processing system from a set of 10 profiles exhibited in Figure A1-3. The values of d_{PF} were found from Eq. A1-1; P was calculated from Eqs. A1-3, A1-4, and A1-5. This gave rise to a set of three values of d_{PF} for each profile, which was arithmetically averaged. The value of R_F for each of the profiles was determined independently according to the procedure outlined in the preceding paragraph. The following empirical relationship has been derived from the values of R_F for all the profiles;

$$d_{PF} = 0.823 + 0.167R_F \qquad \text{(A1-8)}$$

This relationship is graphically represented in Figure A1-4 for all profiles, A through J, in Figure A1-3. With the availability of Eq. A1-8, the perimeter fractal dimensions of the char particles were recovered from the corresponding values of R_F for these particles.

A1.3 RESULTS AND DISCUSSION

The values of d_{PF} for the char samples obtained with the present method are listed in Table A1-2. They range from 1.042 to 1.095. These values tend to indicate that the boundaries of the char samples are not very rugged. This may be attributable to a low degree of ruggedness of the profile of the feed material, wood chips. A careful examination of the chips in the feed has revealed that they resemble the shape of a slab. Even though the char particles from the downdraft gasifier have undergone substantial shrinkage and fragmentation, they appeared to maintain the shape of a thin slab, thus giving rise to a low degree of ruggedness.

The char particles, having Feret diameters between 0.5 and 2 cm, might have experienced some mechanical breakage by the stirrers or tuyeres (Figure A1-1). Nevertheless, such breakage was probably very slight because of the slow grate rotation maintained throughout the experiment. The highest grate rotation speed recorded in the present work was 12 rph (Table A1-1).

In the present downdraft gasifier, a local temperature of as high as 1000° C was recorded. The char particles that went through the high temperature locale would have

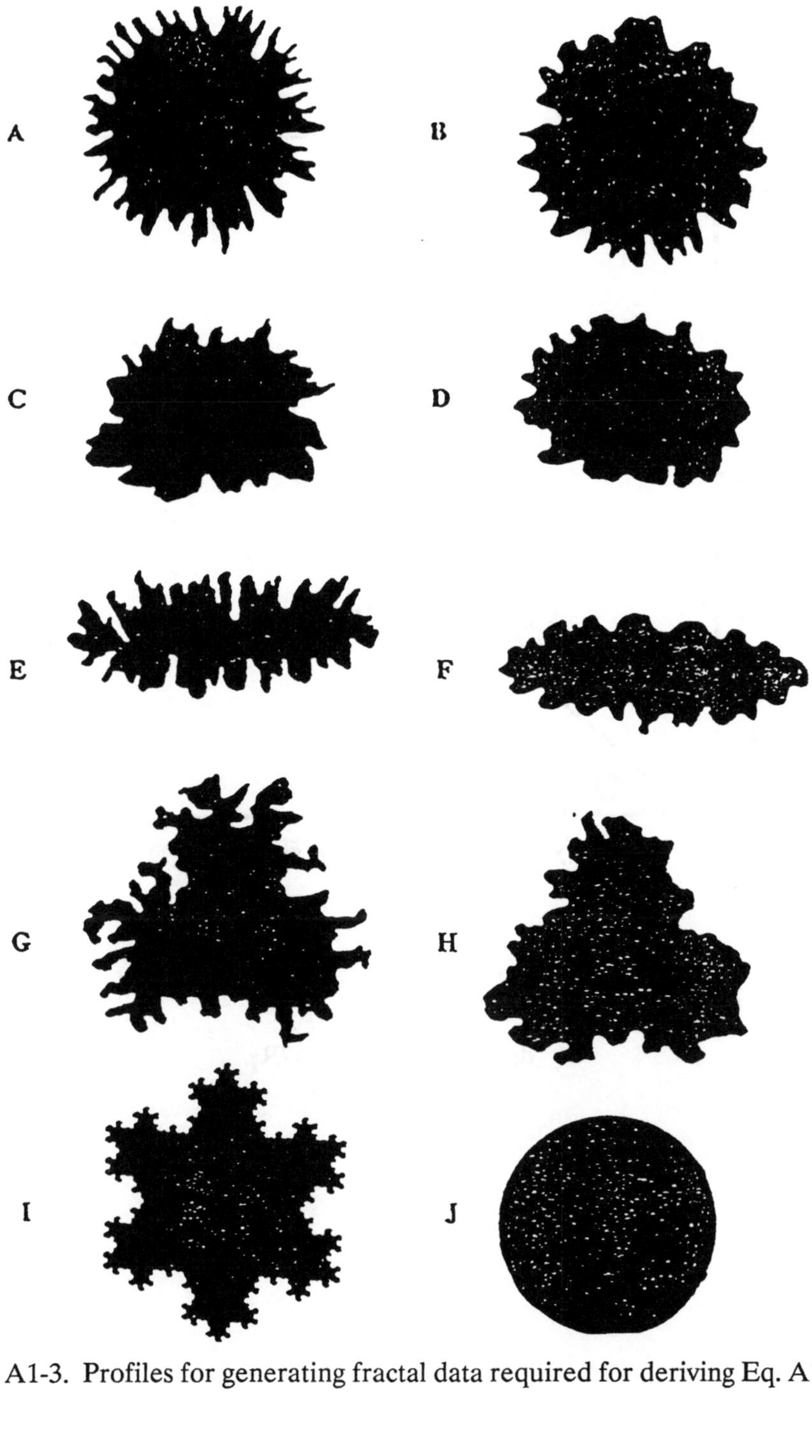

Figure A1-3. Profiles for generating fractal data required for deriving Eq. A1-8.

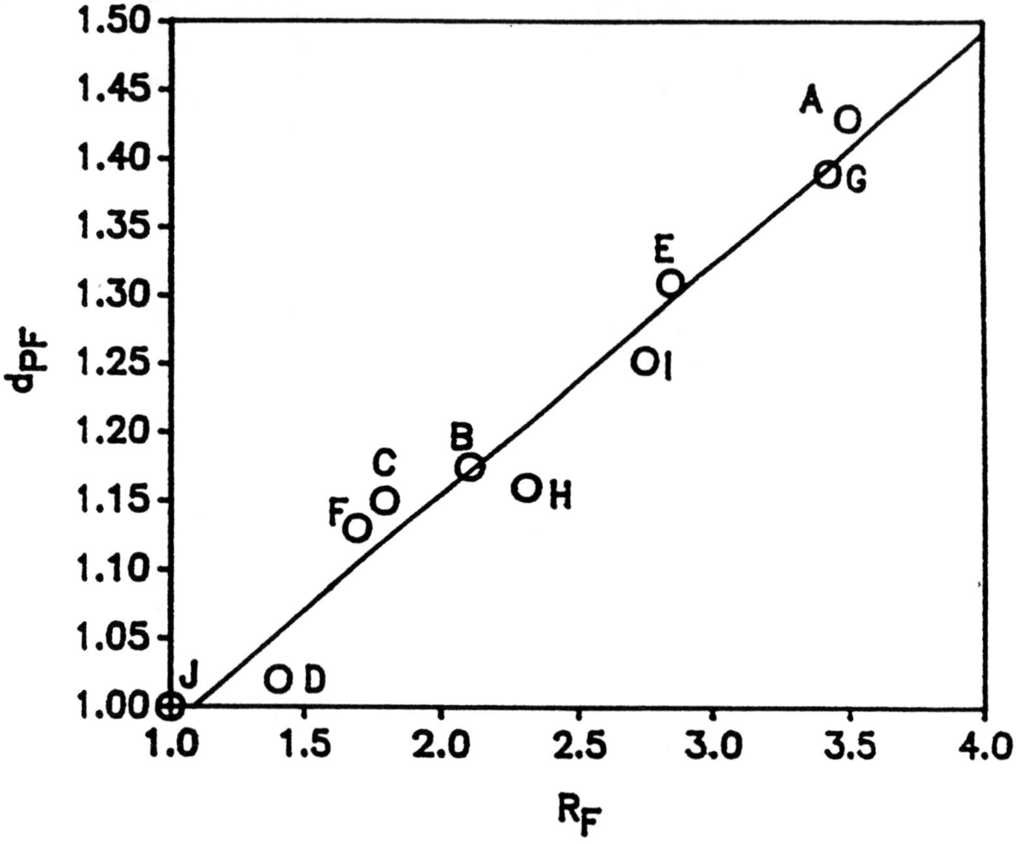

Figure A1-4. Relationship between d_{PF} and R_F derived with the profiles depicted in Figure A1-3:

$$d_{PF} = 0.823 + 0.167 R_F$$
$$R^2 = 0.9880.$$

Table A1-2. Values of d_{PF} of Char Samples Obtained from Various Methods

Run No.	Methods			
	Present	ED	LT	MA
1	1.057	1.066(1.021)	1.061	(1.013)1.021
2	1.087	1.003	1.035	1.033
3	1.062	1.071(1.032)	1.079	1.044(1.026)
4	1.095	1.107(1.024)	1.054	1.001
5	1.042	1.016	1.048	1.203(1.001)
6	1.096	1.140(1.031)	1.060	1.031
7	1.057	1.066(1.021)	1.061(1.013)	1.021
8	1.083	1.001	1.032	1.019

cracked due to thermal stress. However, since the char particles were brittle, they probably tended to crack in such a way that their rather smooth boundaries were preserved; this could be observed visually. Thus, the shapes of wood chips in the feed and char particles resulting from the gasification were almost identical. Specifically, since the boundaries of the chips were fairly smooth, the boundaries of the char particles were also fairly smooth.

The results presented in Table A1-2 appear to suggest that there is no clear cut trend between the values of d_{PF} and the two operating parameters, the grate rotation and the gas fan rotation speeds. No statistical difference test is reported because of insufficient data.

A1.3.1 Comparison with Other Methods

To verify the results obtained in the present work, d_{PF}'s of the char particles were also determined by three well-established methods. For simplicity, these methods are abbreviated as the ED, LT, and MA methods. The ED method is that leading to the evaluation of d_{PF} through Eq. A1-3 [5], the LT method through Eq. A1-4 [81], and the MA method through Eq. A1-5 [11]. The values of d_{PF} obtained by the ED, LT, and MA methods are presented in Table A1-2. These values ranged between 1.001 and 1.203. The values in parentheses correspond to those obtained with λ less than 0.10, which are different from those obtained with λ larger than 0.10; the existence of multifractals is well known [17]. The values of d_{PF} obtained by the present method are clearly within the range of the values obtained by the three established methods, thereby indicating the consistency of the present method.

The present method is analogous to the method developed by Kaye *et al.* [82]. The empirical equation developed in the present work, Eq. A1-8 is almost identical to Eq. A1-6 derived by them. As stated previously, the former has been derived from 10 fractal profiles in Figure A1-3, which have d_{PF}'s ranging between 1.00 and 1.45, through linear regression between d_{PF} and R_F. The resultant expression has a rather high coefficient of determination, R^2, of 0.9880, indicating a satisfactory correlation between them.

A1.3.2 Evaluation of Various Methods

To further assure its validity, the present method was utilized to evaluate the d_{PF} of Koch's Triadic lake, which is a deterministic fractal object with a known theoretical d_{PF}. The resultant value of d_{PF} is compared with those obtained by the three available methods, i.e., the ED, LT, and MA methods (Table A1-3).

The value of d_{PF} estimated by the present work, 1.278, has been calculated from Eq. A1-8 for R_F of 2.85. As described earlier, R_F is the ratio of the number of intersections between the rugged profile and the equally-spaced, parallel lines to the geometrical mean number of intersections between the same set of parallel lines and the largest and smallest circles circumscribing and inscribing, respectively, the rugged profile under inspection. The computed d_{PF} is 1.27% higher than the theoretical value of 1.262 (Table A1-3). This small deviation again appears to partially validate the present method. The corresponding values of d_{PF} of the Koch's Triadic lake estimated by the ED, LT, and MA methods are 1.204, 1.232, and 1.224, respectively, which, in turn, correspond to the percentage deviations of -4.76%, -2.38%, and -3.01%, respectively.

In summary, it can be stated that the present method overestimates the theoretical value while other methods underestimate it. However, all methods yield values of d_{PF} closely approximating the theoretical value.

Table A1-3. Comparisons Between Measured and Theoretical Values of d_{PF} for Koch's Triadic Lake

Methods	Measured d_{PF}	Percent Error (%)
Present	1.278	1.27
ED	1.204	-4.76
LT	1.232	-2.38
MA	1.224	-3.01

A2. APPENDIX 2: Surface Fractal Dimension of Rice Hull-Derived Charcoal from a Fluidized-Bed Reactor

Rice hull, also known as husk, is a by-product of the rice milling industry. This material can be found in abundance in rice producing countries and has posed disposal problems especially in the developing producer countries.

Because of its high energy content ($\sim$ 1500 KJ/Kg), rice hull can be a potential source of energy for the milling processes in the mills themselves, e.g., the generation of steam for parboiled rice, etc. The pyrolysis of rice hull has been proposed as a suitable disposal method, and a series of gasifiers and combustion furnaces have been designed and tested with rice hull as the feed material or fuel (see, e.g., [83]). The major obstacle to the incineration of rice hull under controlled conditions for usable energy has been the flow characteristics of the rice hull; this is associated with the geometry of the material, its light weight and also its high ash content (20% by weight). The gasification of rice hull involves a heterogeneous reaction with a solid reactant in which structural geometry can obviously give rise to a difficulty. Nevertheless, it appears that little has been done to study the gasification of rice hull. For example, the surface area of raw rice hull is scarcely reported; the only reported value is that by Kaupp [83] who determined the apparent surface area of California rice hull to be 3855 m^2/m^3. In his work, Kaupp fitted a sixth-degree polynomial to describe the geometrical bumps on the outer surface of the hull as observed with an electron microscope at a magnification of 25. He traced the circumference with a computer guided image analyzer. The value obtained may be correct for the external surface area of the outer boundary at the magnification employed. Nevertheless, Kaupp has indicated that the major portion of the surface is internal and thus the gas adsorption technique should be adopted to determine and probe all the surface irregularities as well as the internal pores.

Although the gas adsorption method for the surface area measurements may be better than the optical method of Kaupp [83], the values of the surface area measured with different adsorbates yield different results depending on the molecular size of the adsorbate used. In the case of the optical method, the surface area will depend on the scale of the magnification employed. Lately, the fractal dimension has been added to the repertoire of measures to characterize, quantitatively, the intrinsic value of surface irregularities [19]. The fractal dimension is a scale invariant property, characteristic of a particular object irrespective of the yardstick used (see, e.g., [5]). It provides a convenient parameter for comparing surface properties. Recently activated charcoal from coconut shell and various coal particles have been characterized by their fractal dimension (see, e.g., [84]). Avnir *et al.* [85] have reported that the fractal dimension of activated charcoal decreased from 3 to 2 with the increase in the extent of activation. Since the heterogeneous gas-solid reactions encountered in gasification processes involve surface interactions, the surface fractal dimension of a potential solid fuel, which is difficult to use because of its geometrical property, is an important property to characterize.

To follow the development of the available surface area with the progress of the pyrolysis reaction, rice hull from Guyana was devolatilized in a bench-scale, fluidized-bed reactor at different temperatures. The surface areas of the residual char were determined by the gas adsorption method using the BET apparatus. It is anticipated that the results will be utilized to simulate the kinetics of the gasification process of the biomass.

A2.1 THEORETICAL

A fractal surface may be described as that surface in which increasing and similar details are revealed with increased magnification; i.e., the dimension is scale invariant [86]. A surface with high irregularity is characterized by a higher fractal dimension, d_{SF}, which is bounded between the topological dimension of 2 and the Euclidean dimension of 3. For more details readers are referred to sections 2 and 3 of the text.

Two models are available for the determination of the surface fractal dimension by gas adsorption. In the first, the adsorbate is varied and the specific surface area, S, is related to the surface fractal dimension, d_{SF}, by a power law relationship as:

$$S \propto \sigma^{-(d_{SF}-2)/2} \tag{A2-1}$$

where σ is the adsorbate molecular cross-sectional area [19]. Equation A2-1 is another representation of Eq. 3-13 of the text. In the second model, the adsorbent (solid) is varied in particle size while the adsorbate (gas) is fixed. This is in accordance with the fact that dimension can describe a surface and that a similar power law relationship will appear when a larger and larger specimen of the adsorbent is probed by a fixed adsorbate [87]. In this model, the specific surface area, S, is related to d_{SF} as:

$$S \propto R^{d_{SF}-3} \tag{A2-2}$$

where R is the particle radius. Note that derivation of Eq. A2-2 is identical to that of Eq. 3-22. In both of these models, d_{SF} can be recovered from regression analysis relating log S and log σ or log S and log R.

A2.2 EXPERIMENTAL

The experimental procedure comprised devolatilization of rice hull in a fluidized-bed reactor and determination of the surface area of the residual char particles with the BET apparatus employing nitrogen as the adsorbate at 77 K. The rice hull, with properties listed in Tables A2-1 and A2-2, was pyrolyzed in a 2 in I.D bench-scale, fluidized-bed reactor (Figure A2-1) at temperatures of 848, 973 and 1173 K and a retention time of 30 s; the fluidizing gas was nitrogen. The conditions in the fluidized-bed reactor are presented in Table A2-3. A sample of the rice hulls was charged and retrieved from the bed by containing it in an inconel 600 alloy gauze cylinder.

To determine experimentally the surface fractal dimension, d_{SF}, the charcoal produced was sieved to particle sizes ranging from 45 to 1680 μm. Care was taken in screening the samples so as to ensure that self similarity of the surface irregularity was maintained. In other words, the screening was conducted carefully and slowly to prevent any further sample chipping or breakage, thus minimizing the degree of distortion to the original surface areas. The surface areas of the screened particles of different sizes, produced at the same temperature, were evaluated from the BET isotherm.

A2.3 RESULTS AND DISCUSSION

The surface areas of the rice hull derived charcoal, measured as a function of particle size, are listed in Table A2-4. The surface fractal dimensions for the char produced at the various temperatures are presented in Table A2-5. The fractal dimension of the original rice hull prior to devolatilization was 2.42 and those for the charcoal derived at temperatures of 848, 973, and 1173 K in the fluidized-bed reactor have been estimated from Eq. A2-2 to be 2.79, 2.99, and 3.00 respectively. The changes in d_{SF} can be observed through the decrease in the slope of the log-log plots presented in Figures A2-2 through A2-4. A fractal dimension of nearly 3.00, as observed at the higher temperatures, indicates the creation of a highly porous network in the microstructure of the charcoal. This observation is consistent with the negligible changes observed in the BET surface area of the residual char after devolatilization at 973 and 1173 K. Similar results were obtained with BET surface measurements on silica alumina catalyst support, Davidson grade 979, having the same particle size range.

Table A2-1. Physical Properties of As-received Rice Hull from Guyana

Moisture Content	
D.B	12.22% (avg)
W.B	10.86% (avg)
Ash Content	
D.B	21.52% (avg)
W.B	19.15% (avg
Bulk Density	0.0911 g/cc
BET Surface Area*	0.961 m^2/g

* N_2 Adsorption at 77 K measured with BET isotherm.

Table A2-2. Elemental Analysis of As-received Rice Hull

Component	wt(%)
Carbon	37.80
Hydrogen	4.73
Nitrogen	0.46
Oxygen*	35.53
Ash	21.48
Lignin	12.17
Cellulose	45.04
Hemi-cellulose	17.30
Ash	21.54
Moisture	3.96

* By difference

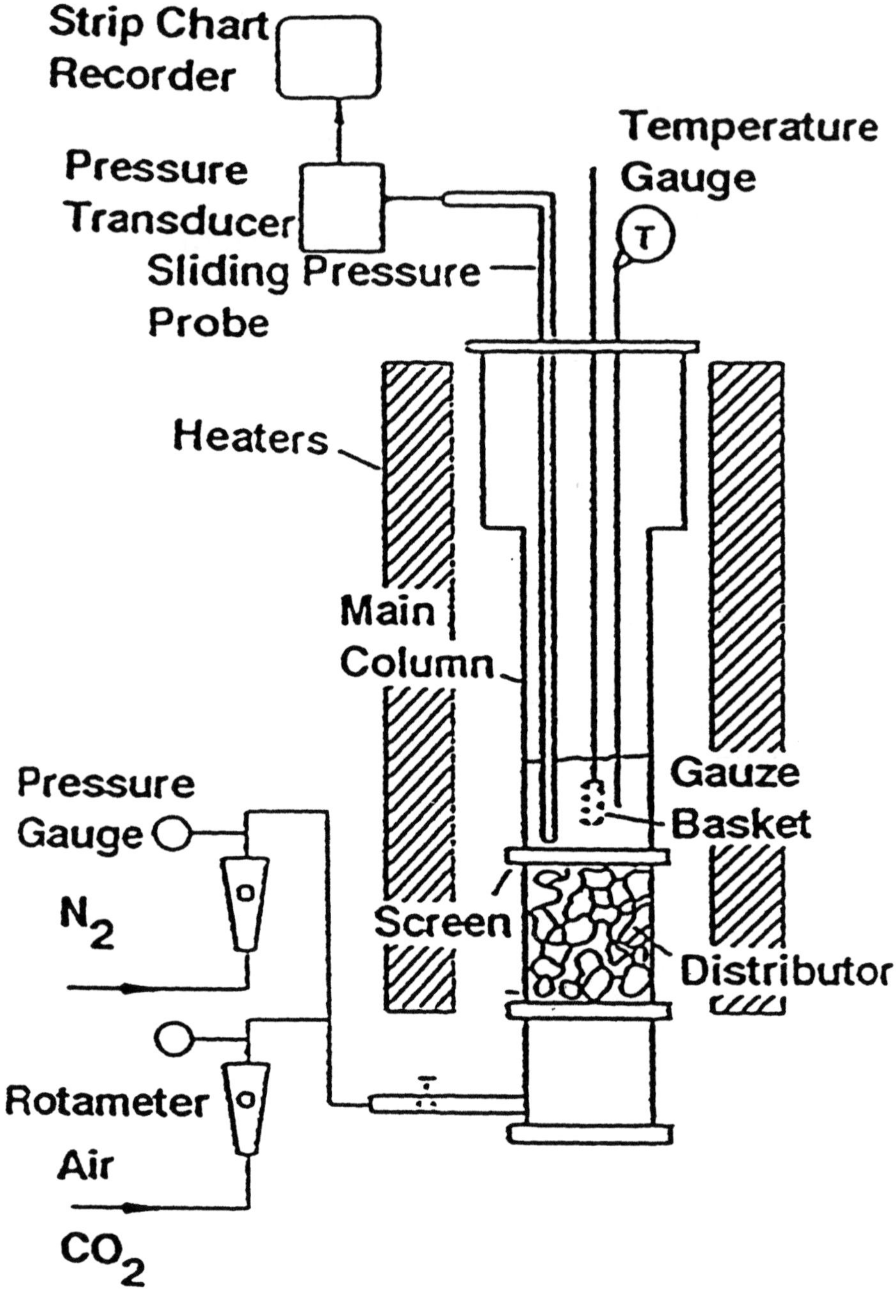

Figure A2-1. Fluidized-bed reactor.

Table A2-3. Experimental Conditions for Rice Hull Devolatilization in the Fluidized-bed Reactor

Rice hull particle size (as-received):	9-12 mm in length
Bed temperature:	848 K, 973 K, 1173 K
Sand/limestone ratio:	8:2
Avg. mixture particle size:	500 μm
Minimum Fluidization Velocity:	
	U_{mf} = 7.5 cm/s @ 848 K
	U_{mf} = 7.0 cm/s @ 973 K
	U_{mf} = 5.0 cm/s @ 1173 K
Superficial gas velocity:	3.2xU_{mf}
Bed diameter:	5.08 cm (2 in)
Static bed height:	7.62 cm (3 in)
Gauze basket diameter:	1.59 cm (5/8 in)
Gauze basket height:	7.62 cm (3 in)
Bed depth in basket:	2.54 cm (1 in)

Table A2-4. Specific Surface Areas of Rice Hull-derived Charcoal in a Fluidized-bed Reactor at Various Temperatures and Char Particle Size

Size Range (μm)	Surface Area (m^2/g)		
	848 K	973 K	1173 K
Original char size before sieving	50.00	185.01	192.86
After sieving			
840 - 1680	33.28	157.97	169.78
590 - 840	98.28	180.29	200.73
420 - 590	93.07	175.42	188.14
250 - 420	120.49	194.50	181.92
149 - 250	119.23	190.88	149.33
45 - 125	126.71	170.03	180.25

Table A2-5. Surface Fractal Dimension of Rice Hull and its Char (d_{SF})

Dev. Temp. (K)	d_{SF} (-)
Orig. rice hull	2.42
848	2.79
973	2.99
1173	3.00

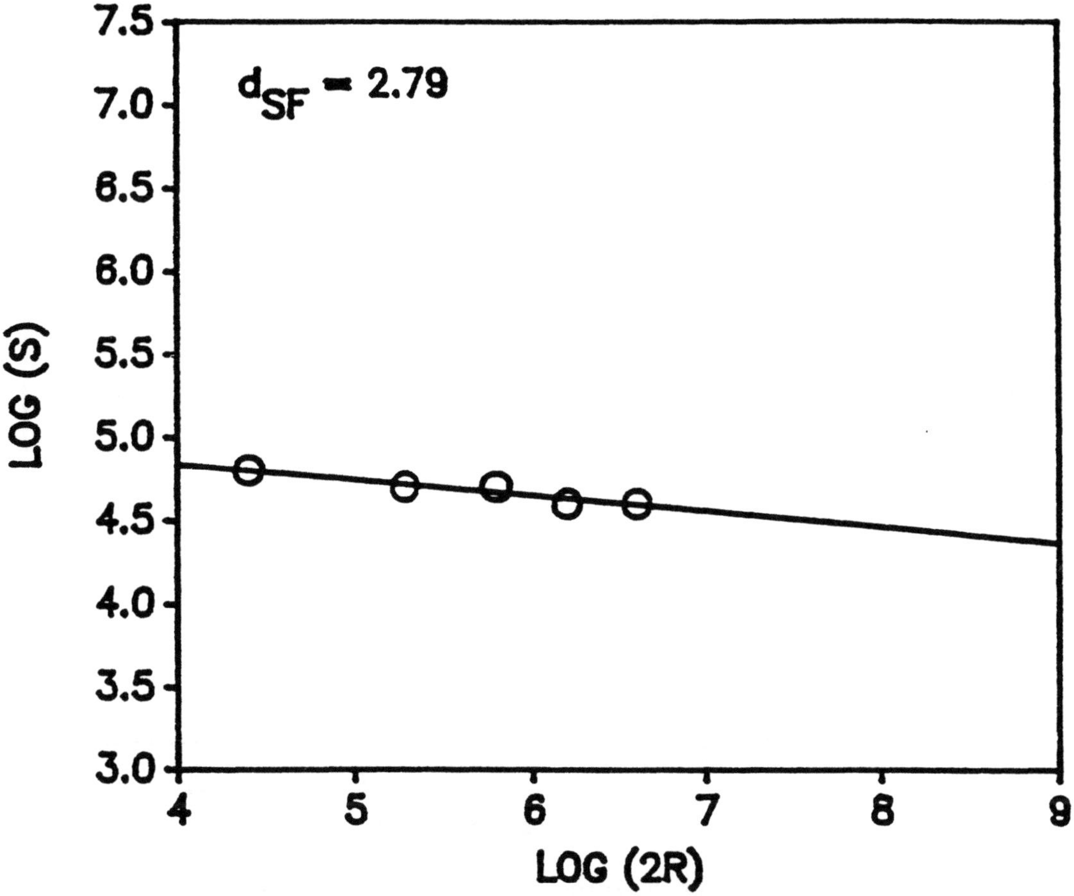

Figure A2-2. Determination of the surface fractal dimension of charcoal derived from rice hull at 848 K.

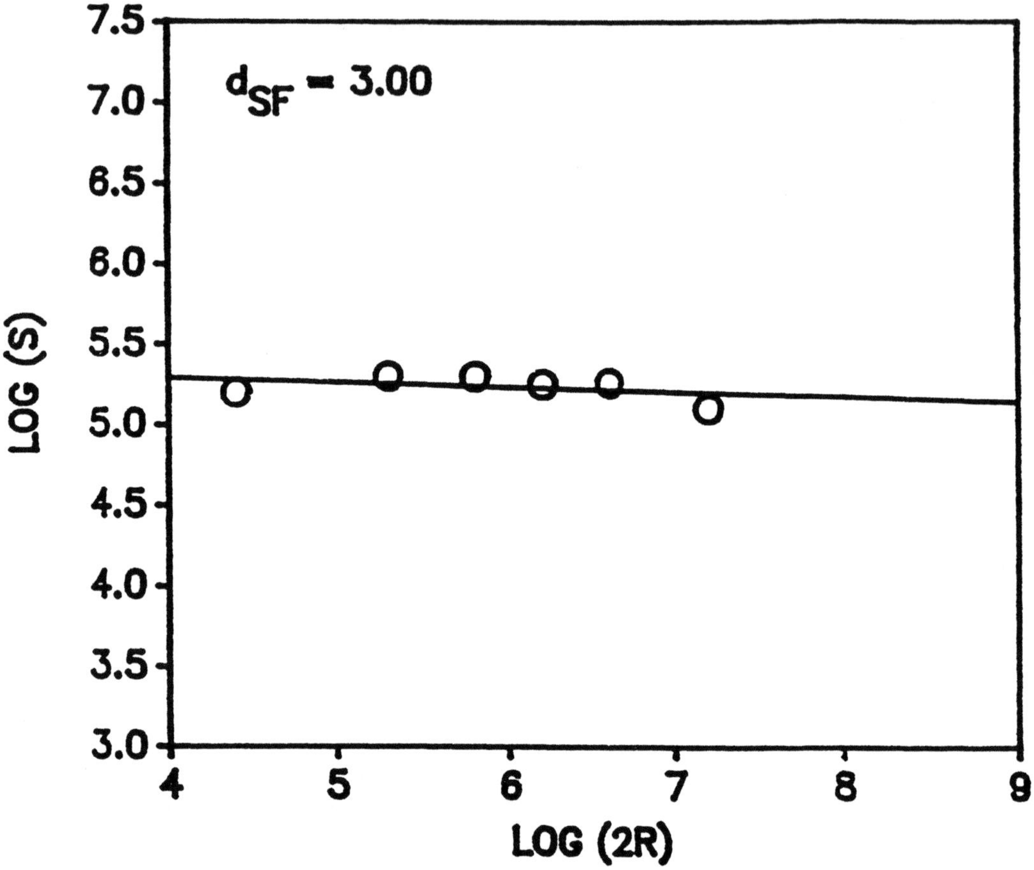

Figure A2-3. Determination of the surface fractal dimension of charcoal derived from rice hull at 973 K.

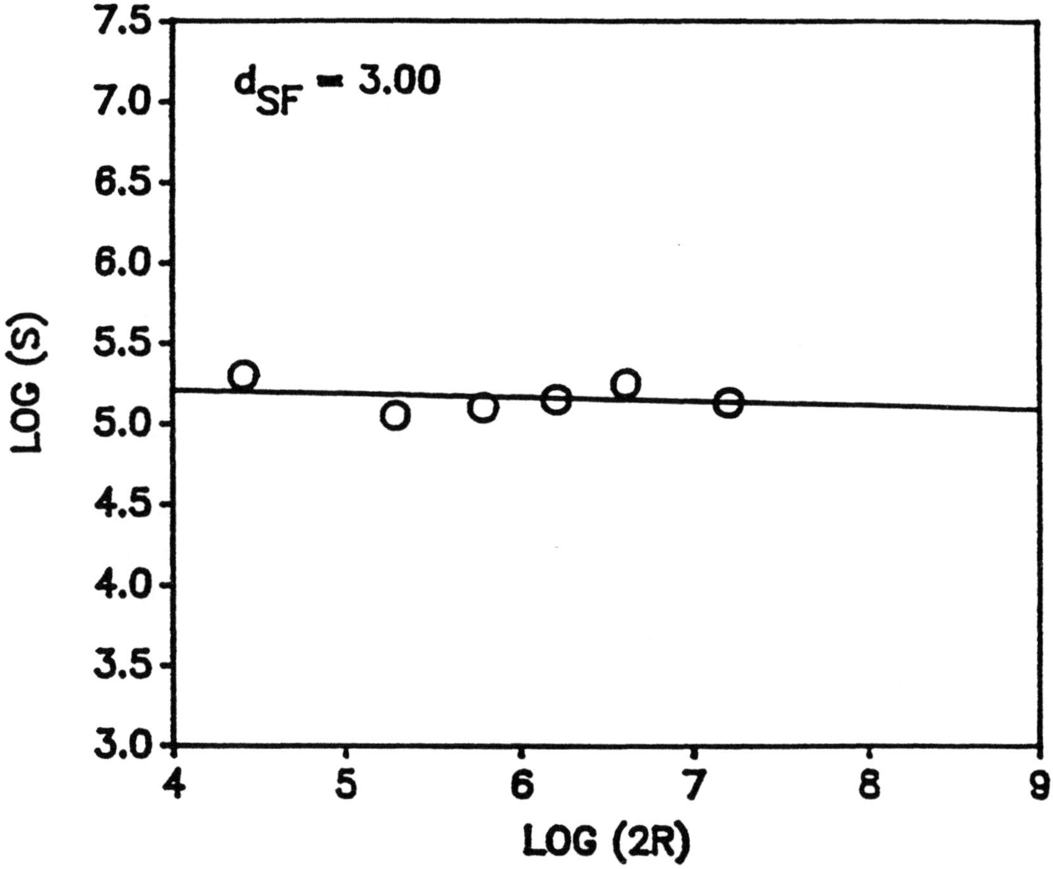

Figure A2-4. Determination of the surface fractal dimension of charcoal derived from rice hull at 1173 K.

The present results indicate that the structure of char generated from rice hull by pyrolysis is sensitive to the temperature of pyrolysis under the fluidized-bed conditions studied. This observation is not surprising since it has been established by Jiang *et al.* [88] that with a relatively low heating rate typical of TGA conditions, specific surface areas increased with increasing devolatilization temperature up to 873-973 K and decreased thereafter. The surface fractal dimension, d_{SF}, can be related to the distribution of the active sites which, in catalytic reactions, have been found to be invariant to scale changes in some cases (see, e.g., [87,89]). The values of d_{SF} recovered for the rice hull charcoal is comparable to that reported for coconut shell charcoal. Avnir *et al.* [85] have reported d_{SF} as 2.30-2.54 for slightly porous coconut charcoal and as 2.67±0.16 whereas a fractal dimension of 2.04±0.04 for porous coconut charcoal, has been reported for active nonporous coconut charcoal. These comparisons show that for rice hull, a highly porous microstructure is already developed at a temperature of 848 K at a relatively short retention time of 30 s in a fluidized bed. This conclusion was substantiated when the pore structure of the charcoal produced was observed by SEM and an image analyzer. The average pore area increased from $4.86/10^{-9}$ m^2 at 848 K to $9.20/10^{-9}$ m^2 at 1173 K for the retention time of 30 s. It is probably possible to simulate the porous network of the microstructure at various temperatures of the pyrolysis reaction with the experimentally determined d_{SF} (see, e.g., [90]).

A3. APPENDIX 3: Fractal Analysis of Pressure Fluctuations in Multiphase Flow Systems

Flow of blood, drift of clouds in the atmosphere, fluidization, pneumatic transport of granular solids, boiling liquids, and, of course, the free flow of champagne are only a few examples of multiphase flow. For an exhaustive review of various multiphase flow systems readers are referred to the handbook edited by Hestroni [91]. Bubble columns and fluidized beds have emerged in recent years as two of the most promising devices for operations involving multiphase flow accompanied, gas-liquid and gas-liquid-solid reactions. Such devices are of considerable industrial importance as evidenced in their wide application for chemical, petrochemical and biochemical processing. Coal gasification and liquefaction, fermentation, waste water treatment, distillation and absorption are examples of processes where bubble columns and fluidized beds have played major roles (see, e.g., [92,93]). Convective boiling, either natural or forced, occurs in numerous industrial systems including thermal power plants and nuclear reactors; studies of boiling abound in the literature (see, e.g., [94,95,96]). Instability in boiling systems has been investigated extensively since it is difficult to predict and tends to persist; it may lead to severe local overheating and damage. Publications on various multiphase flow systems, the majority of which are concerned mainly with the deterministic or global properties of such systems, were reviewed extensively (see, e.g., [91,92,95,97,98,99,100]). Thus, it is essential that we thoroughly understand static and dynamic characteristics of multiphase flow systems if they are to be adopted as processing equipment. Such characteristics manifest themselves as fluctuations of their properties, including density, pressure and composition.

Numerous multiphase flow systems lend themselves to a stochastic-based description due to their inherent complexities and concomitant fluctuations in their behavior. Bubble columns and gas-solid fluidized beds behave highly randomly or stochastically as a result of the influence of a variety of phenomena, including jetting and bubbling of phases, breakage and coalescence of bubbles, motion of fluidized particles; pressure fluctuations are direct consequences of these phenomena. Dynamic instabilities in boiling loops are rather complicated phenomena arising from various mechanisms, including density and pressure wave propagations, thermodynamic nonequilibria and flow-regime transitions (see, e.g., [96]). Instabilities result in random fluctuations of pressure, temperature and density. Linked to these instabilities are problems of transient behavior. Transient conditions often plague startup and shutdown; they are of greatest concern during emergency operations. The most common transients in boiling loops are pressure surges and water hammer effects. Hence, it is imperative that an in-depth understanding be obtained for the time-response behavior of multiphase flow systems if the goal of proper and safe design of control systems is to be achieved [98].

Inherent randomness of the pressure fluctuations in various multiphase flow systems is well known; it was, therefore, natural that statistical methods and stochastic modeling techniques were employed extensively to investigate such phenomena. Bubble breakage and coalescence and dispersion of phases are known to occur in bubble columns and have been studied with vigor (see, e.g., [92]). These phenomena are particularly amenable to stochastic based formulations (see, e.g., [101]). Studies related to probabilistic modeling of coalescence and breakage in dispersed phase systems have been studied by Valentas and Amundson [102], Ramakrishna and Shah [103], and Bajpai *et al.* [104], among others.

Extensive analytic and experimental research on flow instabilities and periodic oscillations in boiling systems has been carried out (see, e.g., [95,97,105,106,107]). Ishii [96] has noted that high-frequency oscillations, which are associated with the propagation of pressure waves, have been observed in some experiments involving boiling. However, very few researchers have attempted statistical studies of pressure fluctuations occurring in boiling loops.

Whitehead *et al.* [108] have indicated that analyzing the pressure fluctuations in a fluidized bed yields information useful for devising control strategies for the bed. Numerous researchers, including Swinehart [109], Lirag and Littman [110], and Fan *et al.* [111], have attempted statistical studies of pressure fluctuations generated by gas-solid fluidized beds. Stochastic models for fluidized beds have been discussed by Shah *et al.* [112], Ligon and Amundson [113,114], Yutani *et al.* [115] and more recently Neogi *et al.* [116]. Both Yutani *et al.* [115] and Neogi *et al.* [116] have considered that any of the pressure fluctuation signals from gas-solid fluidized beds consists of two components, a periodic component and a random component; the latter is modeled as a continuous-time Markov process.

The power of the statistical method and stochastic modeling has been amply demonstrated in analyzing the random pressure fluctuations in multiphase flow systems. Nevertheless, no generalized methodology has been developed to analyze stochastically the transient behavior of such systems.

Hurst [51], while analyzing various hydrological data, has realized that numerous records of observations over time exhibit anomalous characteristics, now known as the "Hurst Phenomenon." Our exhaustive review of the available data and the results of our preliminary experiments have indicated that the time series of pressure fluctuation signals from a convective boiling loop and a fluidized bed exhibit long-term correlation similar to that observed in many hydrological time series. So far the Markov processes, including the Brownian motion, have played dominant roles in modeling the pressure fluctuations in various multiphase flow systems (see, e.g., [115,116]). Nevertheless, they tend to underestimate long term trends, such as the range of "cumulative departure" from the mean as suggested by Hurst [51]. This observation is not surprising; a Markov process has a short-term memory with exponential decay, and Brownian motion comprises independent Gaussian white noise. In contrast, the "fractional Brownian motion" (fBm) model, proposed by Mandelbrot and van Ness [46], is capable of identifying and interpreting long-term persistence or correlation in a time series. It is an extension of the central concept of Brownian motion that has played a significant role in both physics and mathematics; fBm has been one of the major contributions in the field of temporal fractals. This has led us to believe that deeper insights in the stochastic behavior of multiphase flow systems can be gained by resorting to the concept of fractional Brownian motion.

In the present work, pressure fluctuations of three multiphase flow systems, namely a bubble column, a convective boiling loop and a gas-solid fluidized bed, were measured over a range of time by means of pressure probes under various operating conditions. The resultant time series have been analyzed in terms of fBm, or more specifically, in terms of Hurst's Rescaled Range (R/S) analysis. This has yielded the estimates for the so-called Hurst exponent, H, and eventually the local fractal dimension of the time series, $d_{F\ell}$, of pressure fluctuations.

A3.1 THEORETICAL

A fractional Brownian motion, $B_H(t)$, is a single-valued function of one variable, t (usually time). In appearance, it is reminiscent of a mountainous horizon or the fluctuations of an economic variable or the pressure fluctuation signal from a fluidized bed. H relates the incremental change in B_H to the time difference, Δt, through the scaling law

$$\Delta B_H \propto (\Delta t)^H$$

where

$$\Delta B_H = B_H(t_2) - B_H(t_1)$$

and

$$\Delta t = t_2 - t_1$$

In the classical Brownian motion or random walk, the sum of independent increments or steps leads to a variation that scales as the square root of the number of steps or time increments (see, e.g., [48]); thus, $H = 1/2$ corresponds to the trace of a classical Brownian motion. On the other hand, the trace follows a fractional Brownian motion when $H \neq 1/2$ [46].

$$[B(t+s)-B(t)]$$

of the classical Brownian motion has a Gaussian distribution with a variance proportional to the time lag, s, and the increment

$$[B_H(t+s)-B_H(t)],$$

of an fBm has a Gaussian distribution with a variance proportional to s^{2H}, i.e., the 2H power of the time lag, s. Chapter 4 of the text describes the concept of fBm with sufficient vigor.

Summarizing, we may state that a fractional Brownian motion with $H \neq 1/2$ is characterized by three properties, namely, (a) it is a Gaussian process, (b) its increments, $B_H(t+s) - B_H(t)$, constitute a stationary random process with t being a multiple of s, and (c) it is self-affine. The last property implies that scaling time t by r, and the function, $B_H(t)-B_H(0)$, by the ratio, r^H, results in a function whose distribution is identical to the distribution of $B_H(t)-B_H(0)$.

A3.1.1 Discrete-time Fractional Noise

The majority of the records in time obtained in the real world are discrete time observations. $B_H(t)$ can be considered a tool for interpolating a function of discrete time (namely, the cumulative sum of Gaussian random variables) into a function of continuous time. The sequence of increments of $B_H(t)$, i.e., the sequence of values of

$$\Delta B_H(t) = B_H(t+1) - B_H(t)$$

with integer values of time t, is termed "discrete-time fractional noise." For $H=1/2$, $\Delta B_H(t)$ reduces to a discrete-time Gaussian white noise.

A3.1.2 Rescaled Range Analysis

To obtain information about H for a given time series, namely the pressure fluctuation signals from various multiphase systems, we resort to the rescaled range analysis (R/S analysis), which was originally proposed by Hurst [51]. Nevertheless, Mandelbrot and Wallis [52] were the first to apply this method to the determination of the fractal characteristics of a time series; R/S analysis is delineated in chapter 4.

A3.2 EXPERIMENTAL

The experimental facilities for the three systems investigated are outlined. The procedures for operating these facilities and for carrying out the measurements and computations are described.

A3.2.1 Facilities

Bubble Column. Figure A3-1 depicts the experimental facilities. The bubble column consisted of a columnar section and a distributor; the column was fabricated from "Plexiglass" to permit visual observation. The bed's diameter was 0.0508 m and its height was 0.6096 m. The distributor comprised a wire net (opening of 42 mesh or 0.354×10^{-3}m) and an alumina bead packed bed. Water and air constituted the two phases. The physical properties of water and air and the experimental conditions are summarized in Table A3-1. Pressure taps were installed on the wall of the column at three different heights (0.0413 m, 0.130 m, and 0.308 m) above the distributor.

Boiling Loop. Figure A3-2 depicts the closed boiling loop used for our experiment. The main boiling section consisted of a copper tube, 2" (0.0508m) in diameter, with a 2500 watts immersion heater supplying heat from the bottom. The steam was condensed in a heat exchanger and the condensate was recycled back to the boiling section. Pressure taps were installed on the wall of the boiling section as shown in Figure A3-2. The system pressure was varied from 101.43 to 308.43 kPa (0 to 30 psig) with the corresponding temperature ranging from 373 to 406 K (100 to 133° C). Table A3-2 summarizes the operating conditions.

Gas-solid fluidized bed. A schematic diagram of the gas-solid fluidized bed is shown in Figure A3-3. It comprised a bed proper, a distributor, and a plenum column. The bed proper and plenum column were fabricated from "Plexiglass" to permit visual observation. Their heights were 0.61 m (24") and 0.17 m (6.7"), respectively; they had an

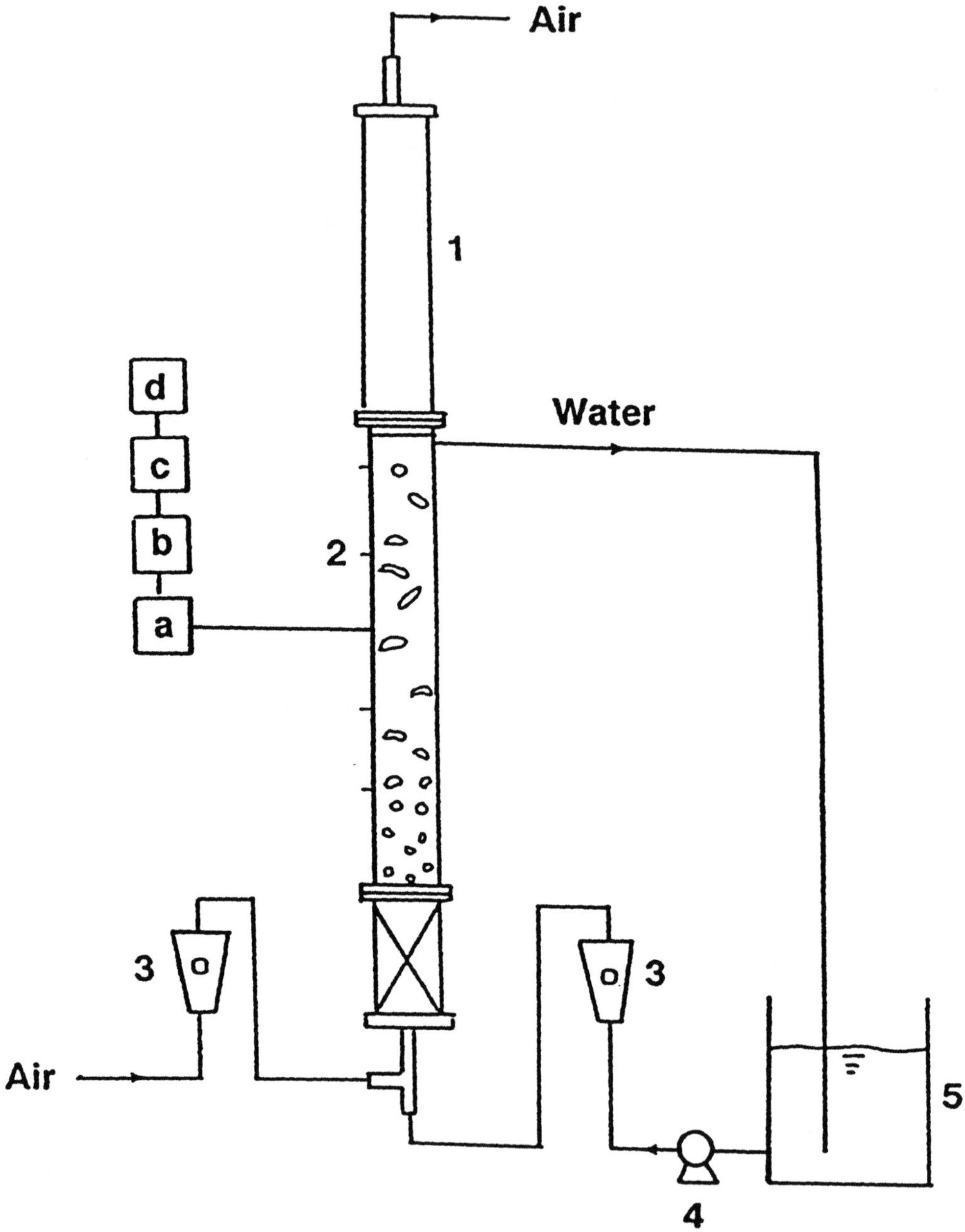

Figure A3-1. Experimental set-up for the bubble column:

1. Main Column
2. Pressure Taps
3. Rotameter
4. Pump
5. Water Reservoir

a. Pressure Transducer
b. Amplifier
c. A/D Converter
d. Micro-computer.

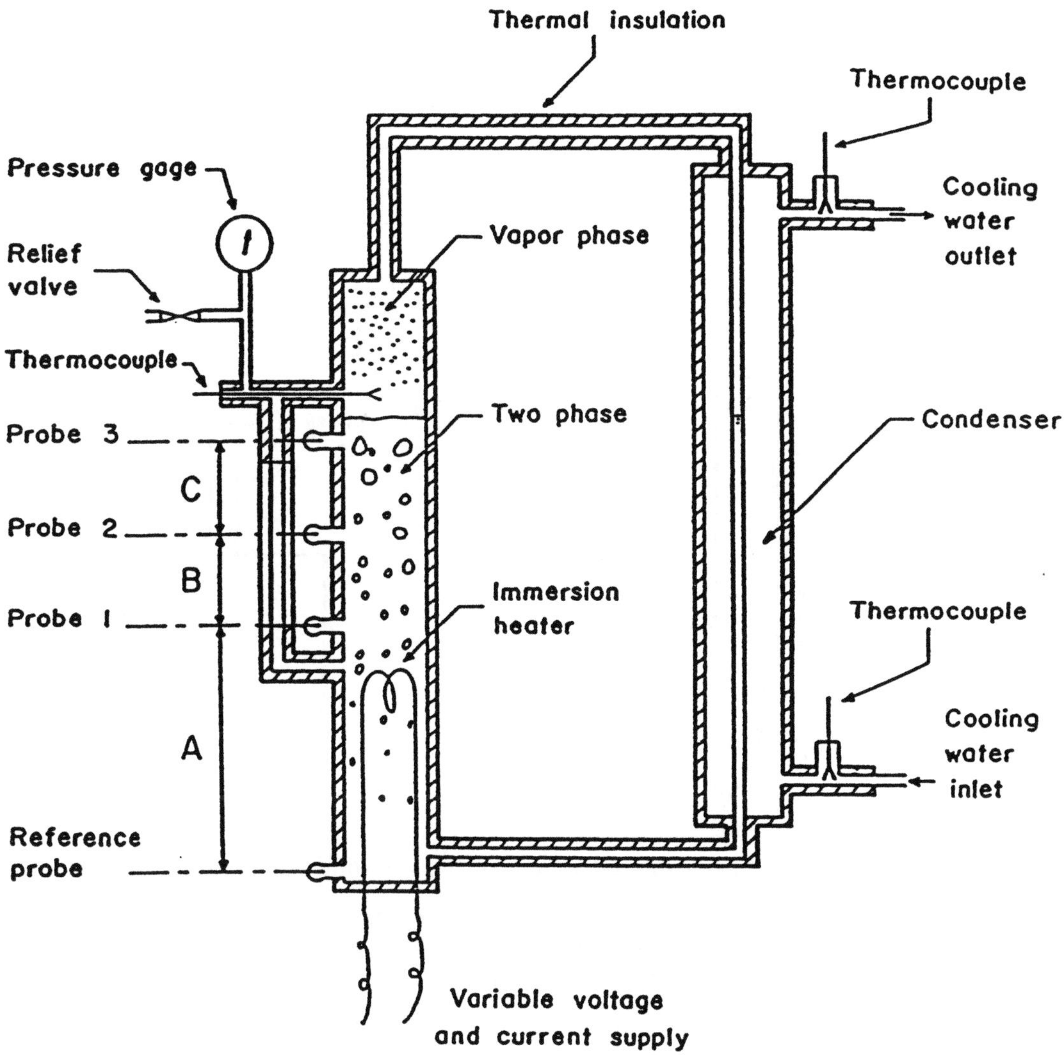

Figure A3-2. Experimental set-up for the boiling loop.

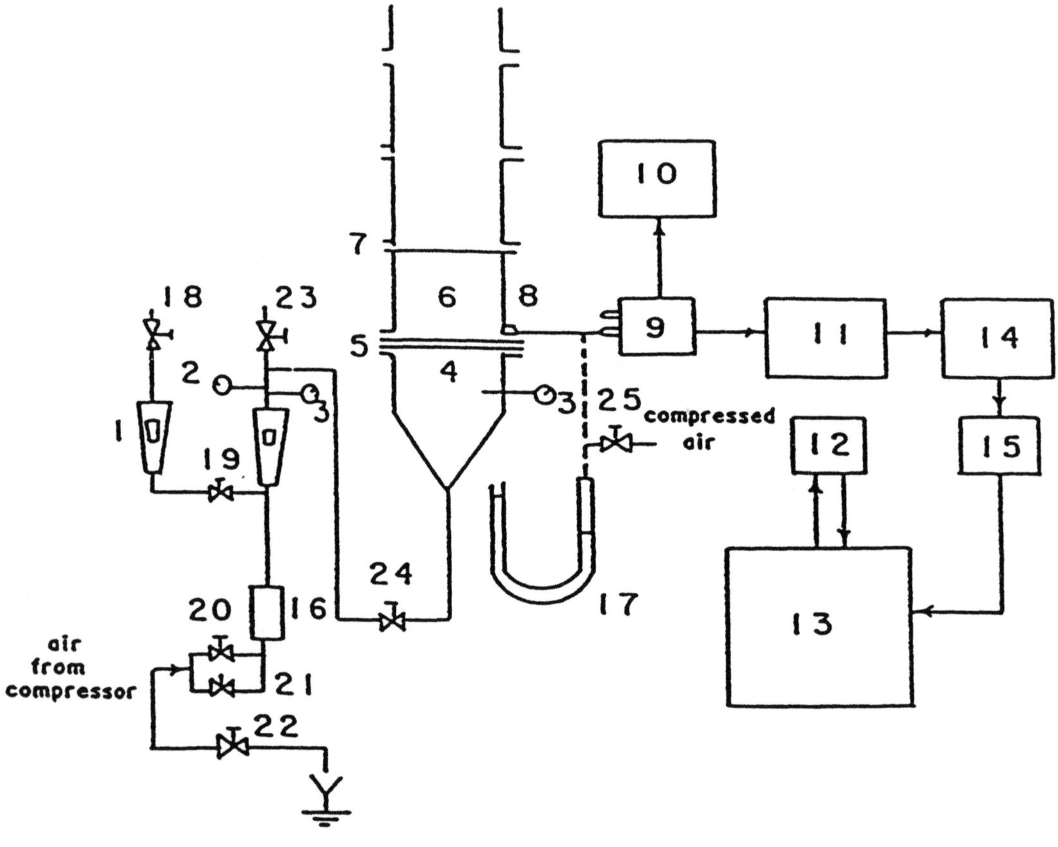

Figure A3-3. Experimental set-up for the gas-solid fluidized bed:

1. Rotameter
2. Pressure gauge
3. Temperature gauge
4. Plenum
5. Distributor
6. Bed
7. Screen
8. Pressure tap
9. Pressure transducer
10. Strip chart recorder
11. Bascom turner recorder
12. SAS package
13. Mainframe
14. IBM PC
15. Modem
16. Drying tube
17. Manometer
18-26. Valves.

Table A3-1. Operating Conditions for the Bubble Column

Water Velocity (m/s)	0	0.056	0.076
Air Velocity (m/s)	0.0129	0.0367	0.0594

Table A3-2. Operating Conditions for the Boiling Loop

System Pressure (kPa)	101.31	156.47	294.35
System Temperature (K)	373	385.63	405.89

identical diameter of 0.203 m (8"). The distributor was perforated aluminum plate, 0.00158 m (1/16") thick and had 164 holes. Sand served as the fluidized particles, and air at 30-35° C and 1 atm(101.31kPa) pressure, the fluidizing medium. The physical properties of the sand and the experimental conditions are summarized in Table A3-3. A pressure tap was installed on the wall of the bed column immediately above the distributor. The inside opening of the tap was covered with a screen to prevent the sand from entering the tap.

Table A3-3. Physical Properties of Sand and Operating Conditions for the Gas-solid Fluidized Bed

Sand Size (m)	0.00084 - 0.00141
Density of Sand (kg/m^3)	2620
Average Sand Diameter (m)	0.000491
Gas Superficial Velocity (m/s)	0.54 - 0.98

A3.2.2 Measurements and Computations

The pressure fluctuations in all three systems were detected by connecting the outside opening of each pressure tap to one of the two input channels of a differential pressure transducer (Enterprise Model CJ3D), which produced an output voltage proportional to the pressure difference between two channels. The remaining channel was exposed to the atmosphere except in the case of boiling loop where the second channel of the transducer was connected to a reference pressure tap located at the bottom of the boiling section. The working capacity of the transducer was ±34.5kPa(±5psi). Signals were processed with the aid of a Nicolet digital oscilloscope, an IBM 386 PC compatible, and a mainframe computer (IBM 3084). The voltage-time signal, corresponding to the pressure-time signal, from the transducer was fed to the recorder at a selected sampling rate; the sampling rates were 0.008 s for the bubble column and the boiling loop and 0.02 s for the gas-solid fluidized bed. A typical sample consisted of 1,984 points. This combination of sampling rate and sample length ensured capturing of the full spectrum of hydrodynamic signals from the three systems investigated; it was typically 25 Hz. The signals were processed off-line.

A3.3 RESULTS AND DISCUSSION

Figures A3-4 through A3-11 depict the signals recorded and the results of processing and analyzing these signals for all the systems investigated. Typical signals recorded from the pressure transducers are traced in Figures A3-4 and A3-5 for the bubble column, in Figure A3-8 for the boiling loop, and in Figure A3-10 for the gas-solid fluidized bed. The superficial velocity of the fluidizing gas is the parameter in Figures A3-4 and A3-10 for the bubble column and gas-solid fluidized bed, respectively; the system pressure is the parameter in Figure A3-8 for the boiling loop. The probe location is the parameter in Figure A3-5 for the bubble column. Notice that each tracing of the recorded signals contains wave-like signals with various frequencies and random signals. In other words, each tracing of the recorded signals appears to comprise a wave-like component and a stochastic component. Also notice that the amplitude of the signals increases with the increase in the superficial velocity for the bubble column and fluidized bed, and it decreases with the increase in the pressure for the boiling loop. The trend for the amplitude to increase with the air flow rate was expected; the bubbles could be visually observed to substantially increase in size when the air flow rate was increased.

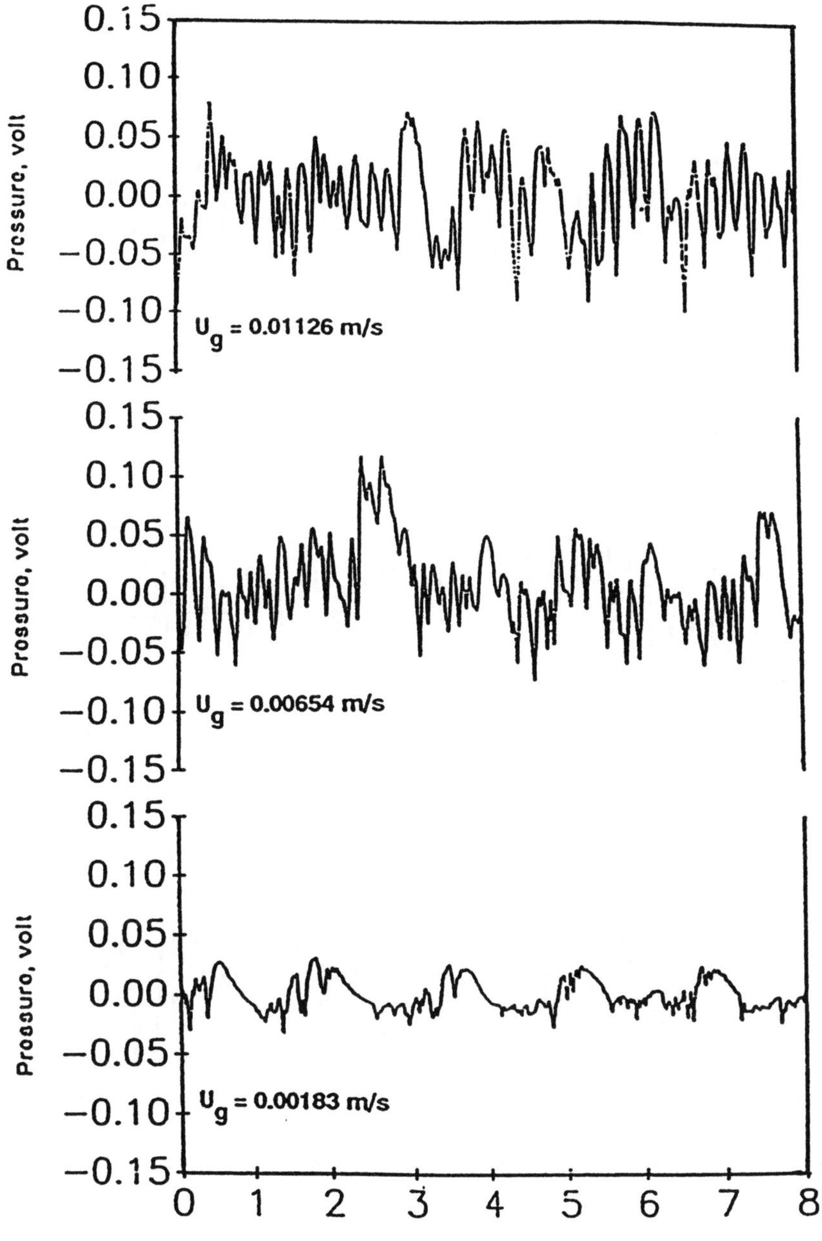

Figure A3-4. Typical pressure fluctuation signals from the bubble column at various superficial gas velocities:

U_ℓ = 0.056 m/s
Probe #2(0.1302m).

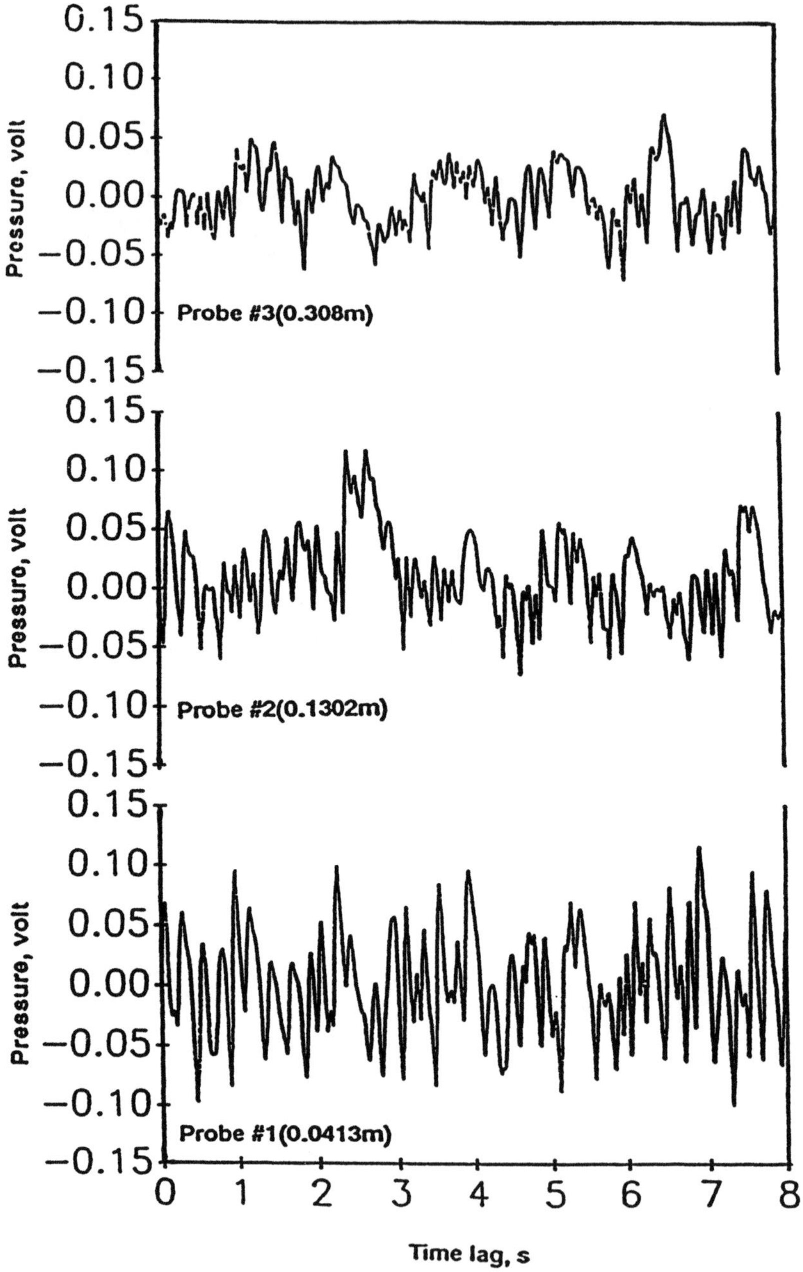

Figure A3-5. Typical pressure fluctuation signals from the bubble column at different probe heights:

$$U_\ell = 0.056 \text{ m/s}$$
$$U_g = 0.00654 \text{ m/s}.$$

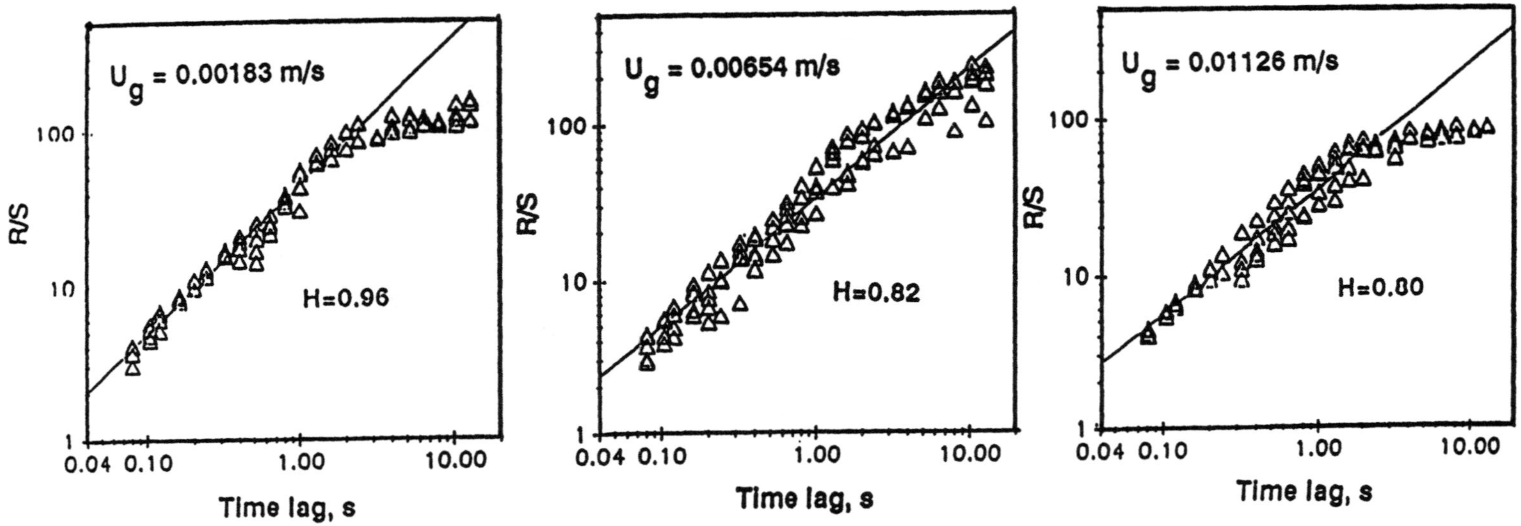

Figure A3-6. Typical Pox diagrams for pressure fluctuation signals from the bubble column at various superficial gas velocities:

U_ℓ = 0.056 m/s
Probe #2(0.1302m).
Note: $d_{FL} = 2 - H$

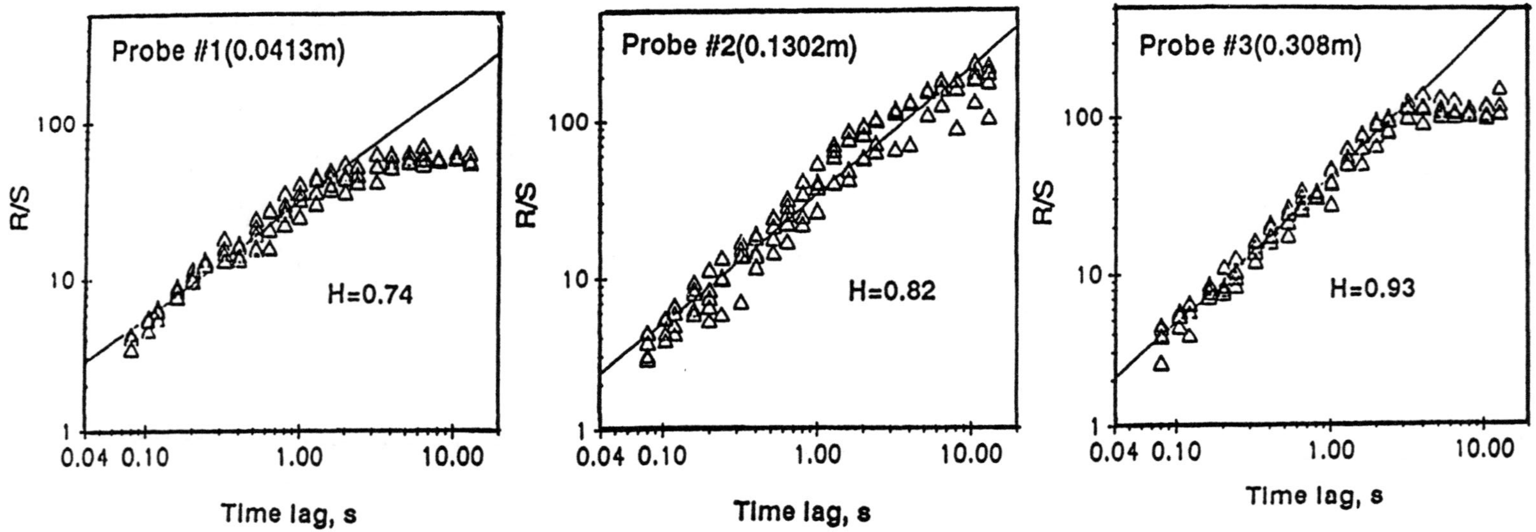

Figure A3-7. Typical Pox diagrams for pressure fluctuation signals from the bubble column at different probe heights:

$U_\ell = 0.056$ m/s
$U_g = 0.00654$ m/s.
Note: $d_{FL} = 2 - H$

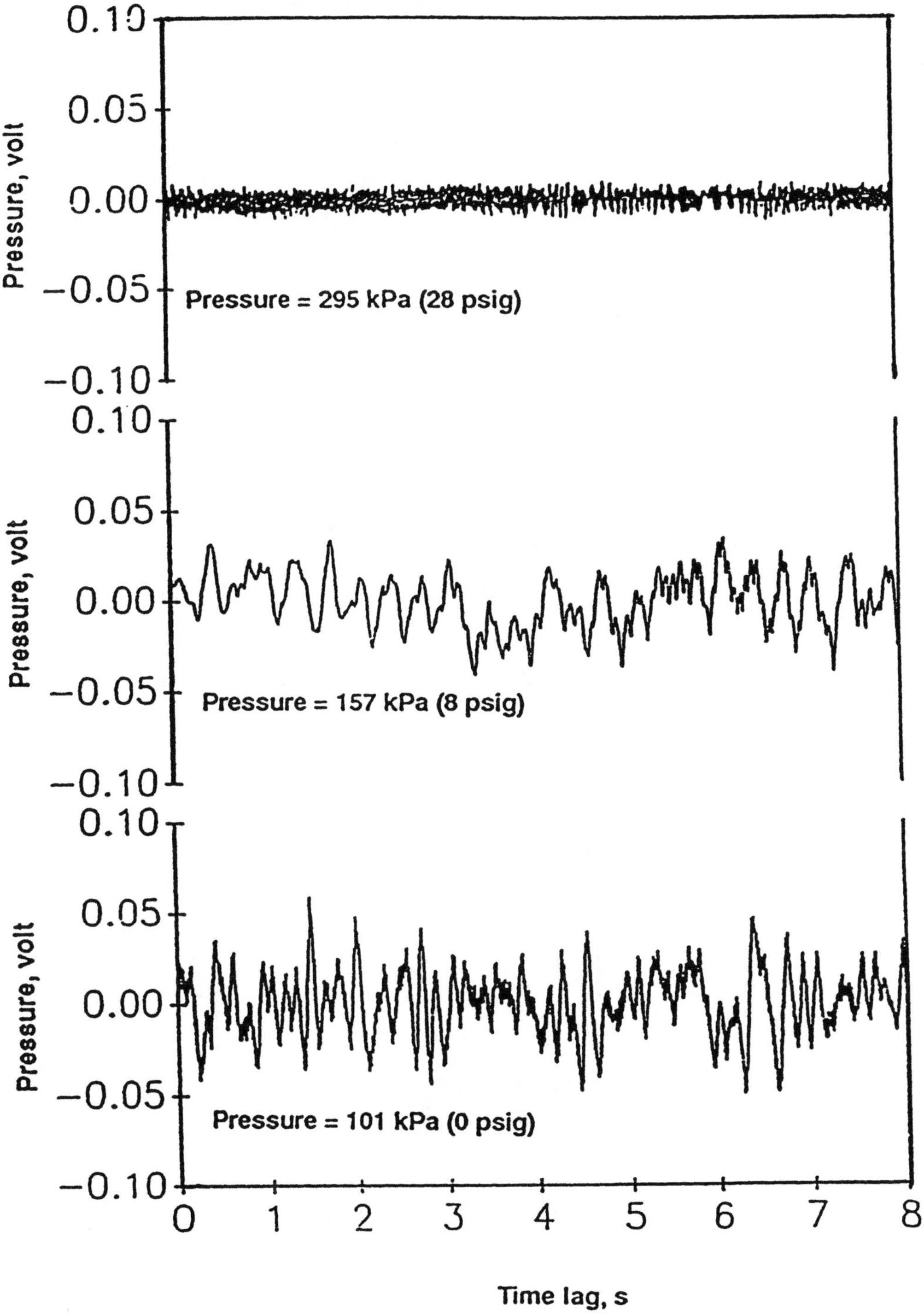

Figure A3-8. Typical pressure fluctuation signals from the boiling loop at different system pressures.

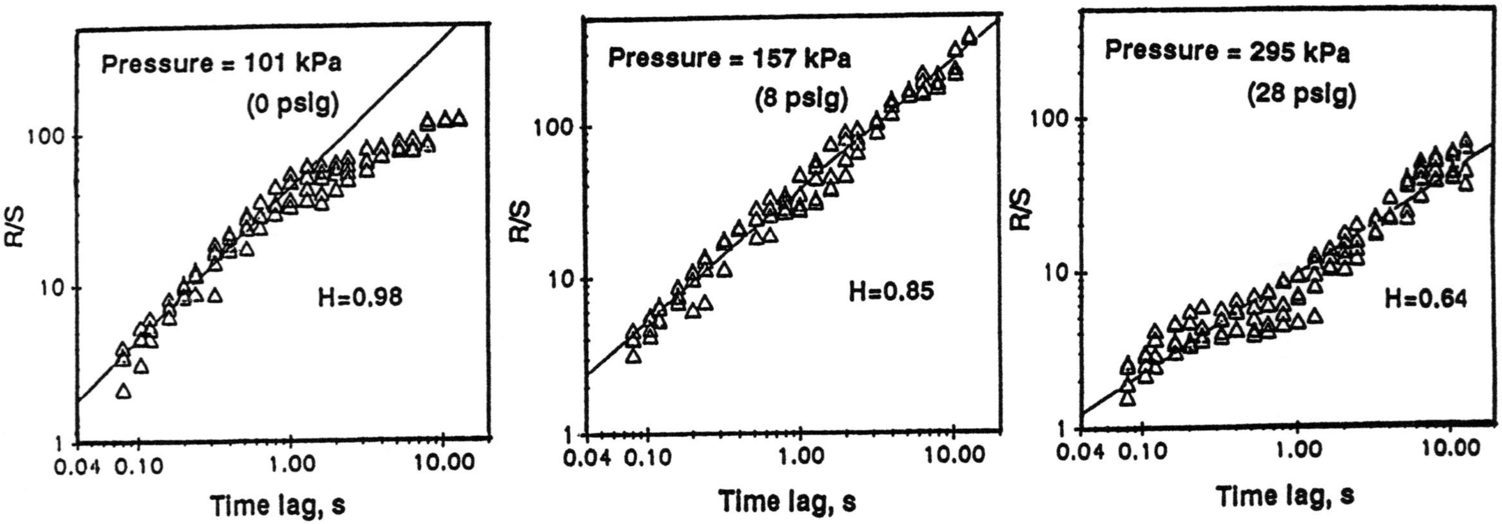

Figure A3-9. Typical Pox diagrams for pressure fluctuation signals from the boiling loop at different system pressures.

Note: $d_{FL} = 2 - H$

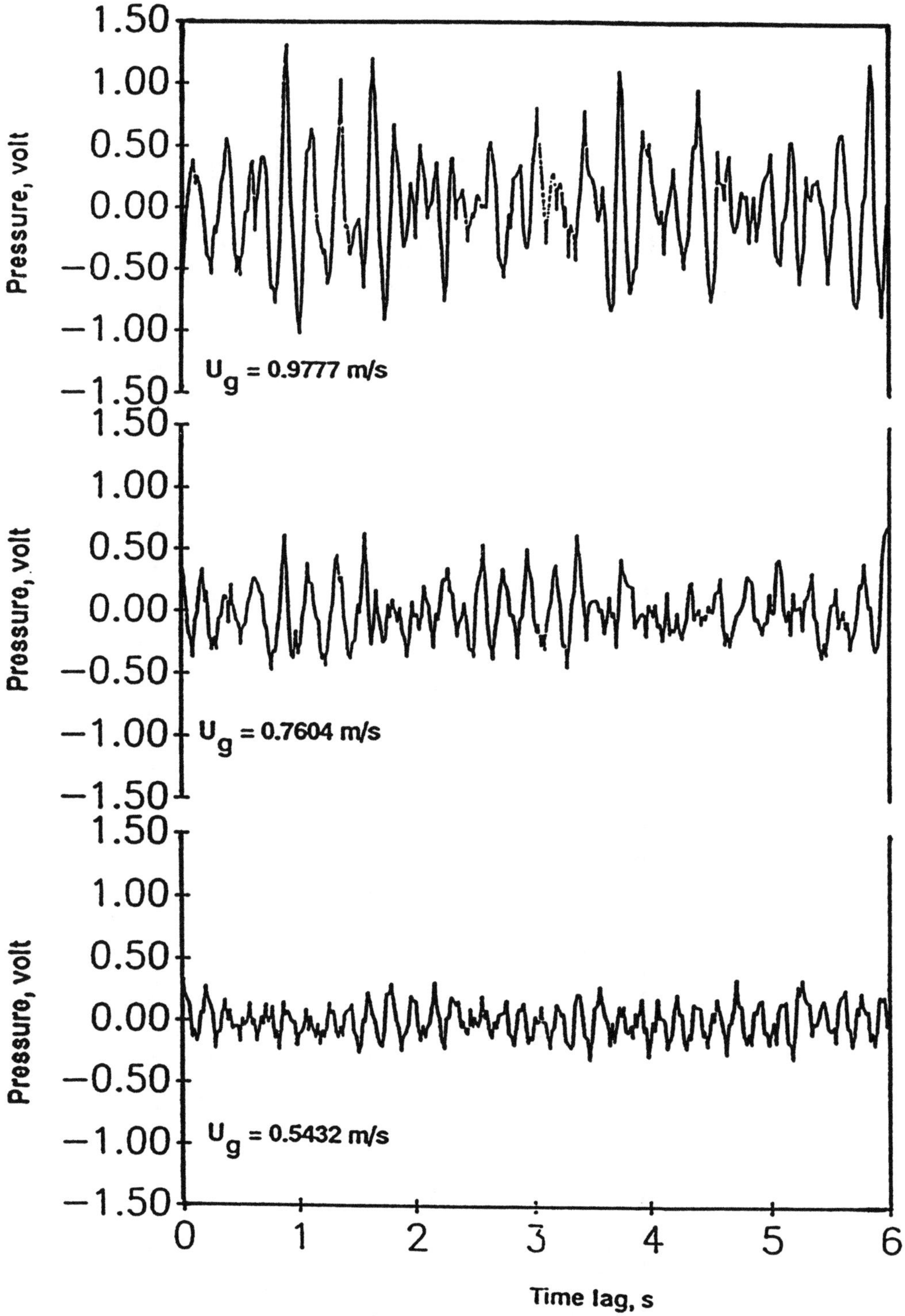

Figure A3-10. Typical pressure fluctuation signals from the gas-solid fluidized bed at various superficial gas velocities.

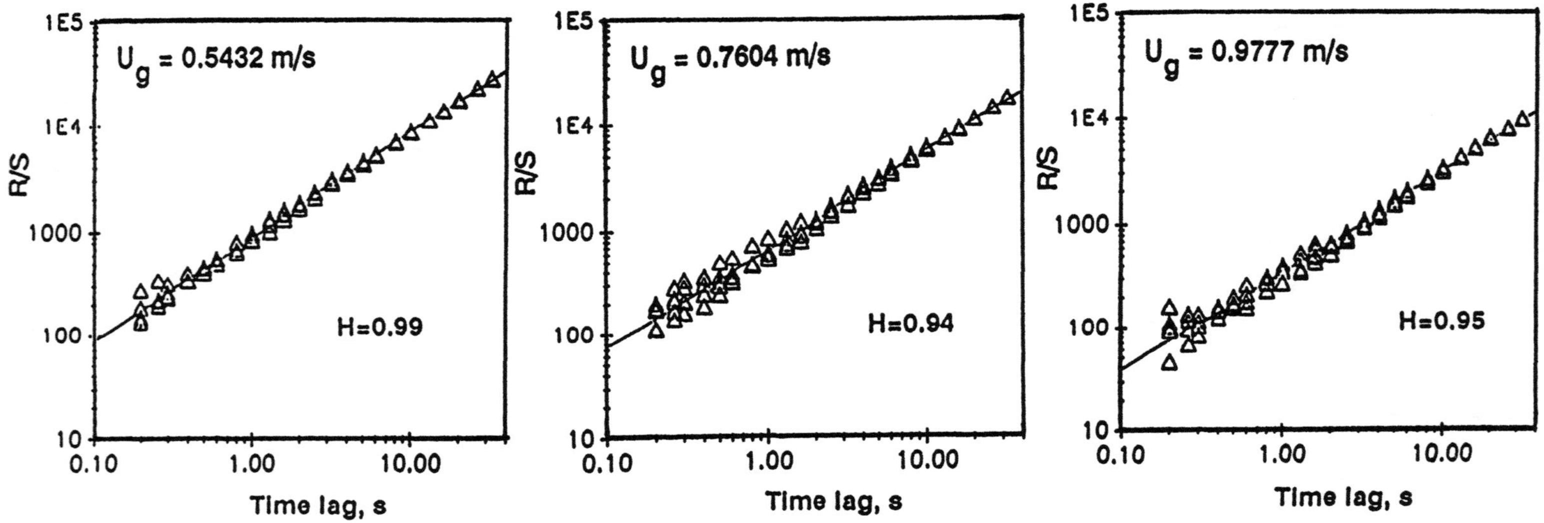

Figure A3-11. Typical Pox diagrams for pressure fluctuation signals from the gas-solid fluidized bed at various superficial gas velocities.

Note: $d_{FL} = 2 - H$

On the other hand, the bubbles in the boiling loop could be observed to decrease in size with the increase in the pressure, thus reducing the amplitude of the recorded signals.

Figures A3-6, A3-7, A3-9, and A3-11 illustrate the log-log plots of R/S against the time lag, s, for the three systems investigated; these plots are known as the "Pox diagrams" [52]. The presence of the five data points at each lag, some of which are overlapping, is due to the fact that rescaled range (R/S) has been computed for each lag by using five different starting times. Figure A3-6 for the bubble column and Figure A3-11 for the gas-solid fluidized bed demonstrate that the Hurst exponent, H, tends to decrease (the local fractal dimension, $d_{F\ell}$, tends to increase) with the increase in the gas velocity, U_g. This could be attributed to the increase in turbulence at higher gas velocities, thus rendering the bubble movements irregular, which, in turn, leads to less persistency in the pressure fluctuation signals. The Hurst exponent, H, decreased ($d_{F\ell}$ increased) with the increase in the system pressure for the boiling loop as evident from Figure A3-9; the increase in the system pressure increased the density of steam which led to the decrease in the bubble sizes rendering bubble movements more random and less persistent. With other operating conditions kept fixed, H tended to increase ($d_{F\ell}$ tended to decrease) with the probe height for the bubble column as shown in Figure A3-7. This is attributable to the growth in the size of the bubbles through coalescence and expansion as they move up in the bed; consequently, the bubble movements generate more persistent signals near the top than near the bottom of the bed.

Another feature observable in each of the pox diagrams (Figures A3-6 and A3-7 for the bubble column and Figure A3-9 for the boiling loop) is the presence of a break in some of the R/S curves. Through simulated examples, Mandelbrot and Wallis [117] have attributed such a break to the presence of a periodic component. Furthermore, they have pointed out that the break occurs in the R/S curves at the time lag corresponding to the dominant frequency of the periodic component, i.e., a time series consisting of both periodic and stochastic components will produce an R/S curve with a break. The present work has demonstrated that pressure fluctuations from a bubble column produces an R/S curve with breaks under certain operating conditions. Similar results have been obtained with the boiling loop operating under atmospheric pressure. Apparently, R/S analysis not only confirms the presence of a stochastic component following fBm but also reveals the presence of one or more periodic components.

NOMENCLATURE

a	=	exponent of Eq. 3-2
A	=	area measure
A_S	=	surface area of the adsorbent
b	=	radius of the adsorbate molecule
B	=	height of the rectangle
$B_H(t)$	=	fractional Brownian motion
$B(t)$	=	change of the commodity price;
	=	also Brownian motion
C	=	length of the rectangular solid
$C(r)$	=	density-density correlation function;
	=	also correlation integral
d_E	=	Euclidean dimension
d_F	=	fractal dimension
$d_{F\ell}$	=	local fractal dimension
d_{Fs}	=	structural fractal dimension
d_{Ft}	=	textural fractal dimension
d_H	=	Hausdorff-Besicovitch dimension
d_{PF}	=	perimeter fractal dimension
d_S	=	similarity dimension
d_{SF}	=	surface fractal dimension
d_T	=	topological dimension
D	=	width of the rectangular solid;
	=	also diffusivity
D_H	=	fractal diffusivity
E	=	height of the rectangular solid
$f(\cdot)$	=	density function
$F(\cdot)$	=	distribution function
F_D	=	Feret diameter
$h(r^*)$	=	height of the pore of radius r^*
$h(\lambda)$	=	test function
H	=	Hurst exponent
L	=	length measure
$L(t,s)$	=	change of the logarithmic commodity price
L_c	=	characteristic length of the adsorbent particles
L_p	=	perimeter length

$m(r^*)$ = density function for the number of pores of radius r^*
M = number of replacement steps
M_d = measure of a set
$M(r)$ = number of occupied sites
n = number of moles of the adsorbate corresponding to the monolayer formation
N = number of self-similar parts
N_b = total number of adsorbate molecules
p = embedded phase-space dimension
$p(\cdot)$ = probability
$\tilde{p}$ = characteristic function
P = applied pressure in the mercury porosimeter;
= also perimeter of a profile
q = inverse of self-similar ratio
r = similarity ratio;
= also distance separating two sites
r^* = pore radius
R = characteristic length of the adsorbent particle;
= also particle radius
$R(t,s)$ = sequential range
s = number of self-similar parts;
= also time lag
$S(t,s)$ = sequential standard deviation
t = time
V = volume measure;
= also volume of pores
V_p = volume of mercury forced into the pores
$V(r^*)$ = cumulative volume of pores of radius greater than r^*
$W(k)$ = Weierstrass function
z = height of mercury in porosimeter
$Z(t)$ = commodity price

Greek letters

α = exponent defined in Eq. 3-40;
= also index of peakedness

β	=	index of skewness
γ	=	scale factor
δ	=	location parameter;
	=	also dirac delta function
ϵ	=	length of the measuring unit or edge length of a p-dimensional cube
ζ	=	fraction of the structured walk technique
η	=	surface tension
θ	=	angle used for constructing the signature wave form
λ	=	stride length for measuring the perimeter;
	=	also side of a square;
	=	also edge of a cube
	=	also parameter in Eq. 4-43
λ_i	=	i-th Lyapunov exponent
μ	=	mean
ν	=	uniform density;
	=	also correlation dimension
ξ	=	mean value of the exponential distribution
ρ	=	density
σ	=	effective cross-sectional area of the adsorbate;
	=	also standard deviation
τ	=	inter-arrival time;
	=	also delay time
$\tau_{m\ell}$	=	most likely time for the middle event
ϕ	=	contact angle subtended by the mercury with the surface of pores
$\phi(t)$	=	canonical representation of a family of distribution
$\psi(t)$	=	waiting-time distribution

Subscripts

min	=	minimum
max	=	maximum

LITERATURE CITED

1. Mandelbrot, B.B., "The Fractal Geometry of Nature," pp. 1-24, W.H. Freeman, New York (1982).

2. Gleick, J., "Chaos," pp. 81-118, Viking Penguin, New York (1987).

3. Montroll, E.W. and Shlesinger, M.F., "On the Wonderful World of Random Walks," in Nonequilibrium Phenomena II, J.L. Lebowz and E.W. Montroll, Ed., 1-122, North-Holland Physics Publishing, Netherlands (1984).

4. Pfeifer, P.J. and Obert, M., "Fractals: Basic Concepts and Terminology," in The Fractal Approach to Heterogeneous Chemistry, D. Avnir, Ed., pp. 11-52, John Wiley and Sons, Chichester, Great Britain (1989).

5. Mandelbrot, B.B., "The Fractal Geometry of Nature," pp. 25-83, W.H. Freeman, New York (1982).

6. Mandelbrot, B.B., "The Fractal Geometry of Nature," pp. 362, W.H. Freeman, New York (1982).

7. Feder, J., "Fractals," pp. 11-22, Plenum, New York (1988).

8. Barnsley, M., "Fractals Everywhere," pp. 172-206, Academic Press, San Diego, California (1988).

9. Yamaguchi, M., "Chaos and Fractal" (in Japanese), pp. 183-187, Kodansha, Tokyo (1986).

10. Mandelbrot, B.B., "The Fractal Geometry of Nature," pp. 109-146, W.H. Freeman, New York (1982).

11. Kaye, B.H., "A Random walk through Fractal Dimensions," pp. 13-38, Verlagsgesellschaft, Weinheim, Federal Republic of Germany (1989).

12. McWorter Jr., W.A. and Tazelaar, J.M., "Creating Fractals," Byte, 123-132, August (1987).

13. Richardson, L.F., "The Problem of Contiguity: an Appendix of Statistics of Deadly Quarrels," General Systems Yearbook, **6**, 139-187 (1961).

14. Kaye, B.H., Clark, G.G., Leblane, J.E., and Trottier, R.A., "Image Analysis Procedures for Characterizing the Fractal Dimension of Fineparticles," Part. Charact., **4**, 63-66 (1987).

15. Kaye, B.H., "Fractal Dimension and Signature Waveform Characterization of Fineparticle Shape," American Laboratory, April, 55-63 (1988).

16. Kaye, B.H., "Specification of the Ruggedness and/or Texture of a Fine Particle Profile by its Fractal Dimension," Powder Technology, **21**, 1-16 (1978).

17. Kaye, B.H., "Multifractal Description of a Rugged Fineparticle Profile," Part. Charact., **1**, 14-21 (1984).

18. Fan, L.T., Chee, C.S., and Walawender, W.P., "Perimeter Fractal Dimension of Char Particles from a Downdraft Gasifier Through Image Analysis," presented in the Poster Session on Fundamental Research in Interfacial Phenomena at the AIChE Annual Meeting, San Francisco, CA, November 5-10 (1989).

19. Pfeifer, P.J. and Avnir, D., "Chemistry in Noninteger Dimensions Between Two and Three. I. Fractal Theory of Heterogeneous Surfaces," J. Chem. Phys., **79**, 3558-3565 (1983).

20. Avnir, D., Farin, D., and Pfeifer, P.J., "Chemistry in Noninteger Dimensions Between Two and Three. Part II. Fractal Surfaces of Absorbance," J. Chem. Phys., **79**, 3566-3571 (1983).

21. Orr, C., "Application of Mercury Penetration in Material Analysis," Powder Technology, **3**, 117-123 (1970)

22. Ruthven, D.M., "Principles of Adsorption and Adsorption Processes," pp. 5-6, Wiley, New York (1984).

23. Fan, L.T., Boateng, A.A., and Walawender, W.P., "Surface Fractal Dimension of Rice Hull-derived Charcoal from a Fluidized-bed Reactor," presented at the Poster Session on the Fundamental Research in Interfacial Phenomena at the AIChE Annual Meeting, San Francisco, CA, November 5-10 (1989).

24. Herrmann, H.J., "Growth: An Introduction," in On Growth and Form, H.E. Stanley and N. Ostrowsky, Ed., pp. 3-20, Martinus Nijhoff, Netherlands (1986).

25. Meakin, P., "Computer Simulation of Growth and Aggregation Processes," in On Growth and Form, H.E. Stanley and N. Ostrowsky, Ed., pp. 111-135, Martinus Nijhoff, Netherlands (1986).

26. Eden, M., Proc. Fourth Berkeley Symp. on Math. Stat. and Prob., F. Neyman, **IV**, p. 223, Univ. of Calif. Press, Berkeley (1961).

27. Williams, T. and Bjerknes, R., "Stochastic Model for Abnormal Clone Spread through Epithelial Basal Layer," Nature, **236**, 19-21 (1972).

28. Sawada, Y., Ohta, S., Yamazaki, M., and Honjo, H., "Self-Similarity and a Phase-Transition-like Behavior of a Random Growing Structure Governed by a Nonequilibrium Parameter," Phys. Rev., **A26**, 3557-3563 (1982).

29. Dhar, D. and Ramaswamy, R., "Classical Diffusion on Eden Tree," Phys. Rev. Lett., **54**, 1346-1349 (1985).

30. Rickvold, P.A., "Simulation of a Stochastic Model for Cluster Growth on a Square Lattice," Phys. Rev., **A26**, 647-650 (1982).

31. Finegold, L.X., "Cell Membrane Fluidity: molecular modeling of particle aggregations seen in electron microscopy," Biochem. Biophys. Acta, **448**, 393-398 (1976).

32. Witten, T.A. and Sander, L.M., "Diffusion-Limited Aggregation, A Kinetic Critical Phenomenon," Phys. Rev. Lett., **47**, 1400-1403 (1981).

33. Witten, T.A. and Sander, L.M., "Diffusion-Limited Aggregation," Phys. Rev., **B27**, 5686-5697 (1983).

34. Vicsek, T., "Pattern Formation in Diffusion-Limited Aggregation," Phys. Rev. Lett., **53**, 2281-2284 (1984).

35. Patterson, L., "Diffusion-Limited Aggregation and Two-Fluid Displacements in Porous Media," Phys. Rev. Lett., **52**, 1621-1624 (1984).

36. Nittmann, J., Daccord, G., and Stanley, H.E., "Fractal Growth of Viscous Fingers: quantitative characterization of a fluid in stability phenomenon," Nature, **314**, 141-144 (1985).

37. Tang, C., "Diffusion-Limited Aggregation and the Saffman-Taylor Problem," Phys. Rev., **A31**, 1977-1979 (1985).

38. Matsushita, M., Sano, M., Hayakawa, Y., Honjo, H., and Sawada, Y., "Fractal Structures of Zinc Metal Leaves Grown by Electrodeposition," Phys. Rev. Lett., **53**, 286-289 (1984).

39. Elam, W.T., Wolf, S.A., Sprague, J., Gubser, D.V., Van Vechten, D., Barz Jr., G.L., and Meakin, P., "Fractal Aggregated in Sputter-Deposited $NbGe_2$ Films," Phys. Rev. Lett., **54**, 701-703 (1985).

40. Mandelbrot, B.B., "The Variation of Certain Speculative Price," J. of Business (Chicago), **36**, 394-419 (1963).

41. Shlesinger, M.F., "Williams-Watts Dielectric Relaxation: A Fractal Time Stochastic Process," J. Stat. Phys., **36**, 639-645 (1984).

42. Alexander, S.S., "Price Movements in Speculative Markets: Trends of Random Walks?" Industrial Management Review of M.I.T., II, pt. 2, 7-26 (1961).

43. Mandelbrot, B.B., "The Fractal Geometry of Nature," pp. 326-390, W.H. Freeman, New York (1982).

44. Levy, P., "Théorie de l'addition des variables aléatoires," pp. 94-120, Gauthier-Villar, Paris (1954).

45. Lukacs, E., "Characteristic Functions," pp. 113-138, Charles Griffin, London (1970).

46. Mandelbrot, B.B. and van Ness, J.W., "Fractal Brownian Motions, Fractional Noises and Applications," SIAM Review, **10**, 422-437 (1968).

47. Pietgen, H.O. and Saupe, D., Eds., "The Science of Fractal Images," pp. 21-70, Springer-Verlag, New York (1988).

48. Feder, J., Fractals," pp. 149-192, Plenum, New York (1988).

49. Gefen, Y., Aharony, A., and Alexander, S.S., "Anomalous Diffusion on Percolating Clusters," Physics Review Letters, **50**(1), 77-80 (1983).

50. Mandelbrot, B.B., "Self-Affine Fractals and Fractal Dimension," Physica Scripta,

32, 257-260 (1985).

51. Hurst, E.H., "Methods of Using Long-Term Storage in Reservoirs," Proceedings of the Institution of Civil Engineers, **5**(5), Part I, 519-591 (1956).

52. Mandelbrot, B.B. and Wallis, J.R., "Some Long-Run Properties of Geophysical Records," Water Resources Res., **5**, 321-340 (1969).

53. Fan, L.T., Neogi, D., Yashima, M., and Sur, B., "Fractal Analysis of Pressure Fluctuations in Multiphase Flow Systems," presented in the Session on Fractal Concepts in Engineering Problems at the AIChE Annual Meeting, San Francisco, CA, November 5-10 (1989).

54. Fan, L.T., Neogi, D., Yashima, M., and Nassar, R., "Stochastic Analysis of a Three-phase Fluidized-bed Reactor: Fractal Approach," presented in the Session on Fundamentals of Fluidization and Fluid-particle System II at AIChE Annual Meeting, San Francisco, CA, November 5-10 (1989).

55. Shlesinger, M.F., "Fractal Time and 1/f Noise in Complex Systems," in Perspectives in Biological Dynamics and Theoretical Medicine, Koslow, S.H., A.J. Mandill, and M.F. Shlesinger, Ed., pp. 214-228, New York (1987).

56. Hogg, R.V. and Craig, A.T., "Introduction to Mathematical Statistics," pp. 168-169, MacMillan, New York (1978).

57. van Kampen, N.G., "Stochastic Process in Physics and Chemistry," pp. 217-219, North-Holland, Amsterdam, Netherlands (1981).

58. Ross, S.M., "Introduction to Probability Models," pp. 414-415, Academic Press, Florida (1985).

59. Parzen, E., "Stochastic Processes," pp. 139-144, Holden, California (1962).

60. Thompson, J.M.T. and Stewart, H.B., "Nonlinear Dynamics and Chaos," pp. 1-26, John Wiley and Sons, New York (1988).

61. Moon, F.C., "Chaotic Vibrations: An Introduction for Applied Scientists and Engineers," pp. 12-19, John Wiley and Sons, New York (1987).

62. Brecque, Mort La., "Fractal in Physics," Mosaic, **18**(2), 22-41 (1987).

63. Doherty, M.F. and Ottino, J.M., "Chaos in Deterministic Systems: Strange Attractors, Turbulence, and Applications in Chemical Engineering," Chem. Eng. Sci., **43**, 139-183 (1988).

64. Lanford, O.E., "The Strange Attractor Theory of Turbulence," A. Rev. Fluid Mech., **14**, 347-364 (1982).

65. Gleick, J., "Chaos," pp. 119-154, Viking Penguin, New York (1987).

66. Moon, F.C., "Chaotic Vibrations: An Introduction for Applied Scientists and Engineers," pp. 121-260, John Wiley and Sons, New York (1987).

67. Henon, M., "A Two-Dimensional Mapping with a Strange Attractor," Comm.

Math. Phys., **50**, 69-77 (1976).

68. Grassberger, P. and Procaccia, I. "Measuring the Strangeness of Strange Attractors," Physica D, **9**, 189-208 (1983).

69. Osborne, A.R. and Provenzale, A., "Finite Correlation Dimension for Stochastic Systems with Power-Law Spectra," Physica D., **35**, 357-381 (1989).

70. Takens, F., "Detecting Strange Attractors in Turbulence," in Dynamical Systems and Turbulence, D.A. Rand and L.S. Young, Eds. pp. 366-381 (Springer Lecture Notes in Mathematics, Vol. 898), Springer-Verlag, New York (1980).

71. Farmer, J.D., "Chaotic Attractors of an Infinite-Dimensional Dynamical System," Physica D, **4**, 366-393 (1982).

72. Packard, N.H., Crutchfield, J.P., Farmer, J.D., and Shaw, R.S., "Geometry from a Time Series," Phys. Rev. Lett., **45**, 12-16 (1980).

73. Wolf, S.A., Swift, J.B., Swinney, H.L., and Vastano, J.A., "Determining Lyapunov Exponents from a Time Series," Physica D, **16**, 285-317 (1985).

74. Kaplan, J.L. and Yorke, J.A., "Chaotic Behavior of Multidimensional Difference Equations," in Functional Differential Equations and Approximation of Fixed Points, H.O. Peitgen and H.O. Walther (Eds), pp. 228-237 (Springer Lecture Notes in Mathematics, vol. 730). Springer-Verlag: New York, Heidelberg, and Berlin (1979).

75. Nicolis, C., "Climatic Predictability and Dynamical Systems," in Proceedings of the NATO Advanced Study Institute on Irreversible Phenomena and Dynamical System Analysis in Geosciences, Crete-Greece (1985), NATO ASI Ser., Ser. C, **192**, pp. 321-354, D. Reidal. Hingham., Massachusetts (1987).

76. Brecque, Mort La., "Fractal Applications," Mosaic, **17**(4), 34-48 (1986/87).

77. Walawender, W.P., Chee, C.S., and Fan, L.T., "Operating Parameters Influencing Downdraft Gasifier Performance," in Symposium Papers - Energy From Biomass and Wastes XI, pp. 411-445, Institute of Gas Technology, Chicago (1988).

78. Walawender, W.P., Chee, C.S., and Geyer, W.A., "Influence of Tree Species and Wood Deterioration on Downdraft Gasifier Performance," Biomass, **17**, 51-64 (1988).

79. Walawender, W.P., Chern, S.M., and Fan, L.T., "Influence of Operating Parameters on the Performance of a Wood-Fed Downdraft Gasifier," in Symposium Papers - Energy From Biomass and Wastes XI, pp. 607-627, Institute of Gas Technology, Chicago (1987).

80. Avnir, D., "Fractal Aspects of Materials: Metals and Catalyst Surfaces, Powder and Aggregates," presented at the Annual Meeting of the Materials Research Society, November 26-27 (1984).

81. Kaye, B.H., and Clark, G.G., "Fractal Description of Extra Terrestrial Fineparticles," Part. Charac., **2**, 143-148 (1985).

82. Kaye, B.H., Trottier, R.A., and Clark, G.G., "Line Intersect Frequency for Characterizing the Structure of Branching Structures and Other Fractal Boundaries," presented at the 20th Annual Meeting of the Fine Particle Society, Boston, August 21-25 (1989).

83. Kaupp, A., "Gasification of Rice Hulls, Theory and Praxis," Friedr. Vieweg & Sohn, W. Germany (1984).

84. Freisen, W.I., and Mikula, R.J., "Fractal Dimension of Coal Particles," J. Colloid and Interface Sci., **120**, 263-271 (1987).

85. Avnir, D., Farin, D., and Pfeifer, P.J., "Surface Geometric Irregularity of Particulate Materials: A Fractal Approach," J. Colloid and Interface Sci., **103**, 112-123 (1985).

86. Fairbridge, C., Ng, S.H., and Palmer, A.D., "Fractal Analysis of Gas Adsorption on Syncrude Coke," J. Fuel, **65**, 1759-1762 (1986).

87. Farin D. and Avnir, D., "The Reaction Dimension in Catalysis on Dispersed Metals," J. Am. Chem. Soc., **110**, 2039-2045 (1988).

88. Jiang, W.D., Lee, In-Chul, and Yang, R.Y.K., "Microstructural Variations of Lignite, Subbituminous and Bituminous Coals and Their High Temperature Chars," J. Fuel Processing Tech., **18**, 1-23 (1988).

89. Farin, D. and Avnir, D. "The Fractal Nature of Molecular-Surface Chemical Activities and Physical Interactions in Porous Materials," In Characterization of Porous Solids, K.K. Unger ed., Elsvier Publ. Amsterdam (1988).

90. Sahimi, M. and Tsotsis, T.T., "Statistical Modeling of Gas-Solid Reaction with Pore Volume Growth; Kinetic Regime," Chem. Eng. Sci., **43**, 113-121 (1988).

91. Hestroni, G., Editor-in-Chief, "Handbook of Multiphase Systems," Hemisphere, Washington (1982).

92. Shah, Y.T. and Deckwer, W.D., "Hydrodynamics of Bubble Columns," Handbook of Fluids in Motion, N.P. Cheremisinoff and R. Gupta, Eds., pp. 583-620, Ann Arbor Science, Ann Arbor, MI (1983).

93. Wen, C.Y. and Chen, L.H., "Flow Modeling Concepts of Fluidized Beds," Handbook of Fluids in Motion, N.P. Cheremisinoff and R. Gupta, Eds., pp. 665-690, Ann Arbor Science, Ann Arbor, MI (1983).

94. Fraas, A.P. and Ozisik, M.N., "Heat Exchanger Design," pp. 71-97, John Wiley and Sons, New York (1965).

95. Bouré, J., Bergles, A.E., and Tong, L.S., "Review of Two Phase Flow Instability," Nucl. Eng. Design, **25**, 165-192 (1973).

96. Ishii, M., "Wave Phenomena and Two-phase Flow Instabilities," Handbook of Multiphase Systems, G. Hestroni, Ed., Chapter 2, pp. 95-136, Hemisphere, Washington (1982).

97. Bergles, A.E., "Review of Instabilities in Two-phase Systems," Proceedings of

NATO Advanced Study Institute, Istambul, **1**, 383-422 (1976).

98. Cheremisinoff, N.P., "Fundamentals of Gas-Liquid Flows," Handbook of Fluids in Motion, N.P. Cheremisinoff and R. Gupta, Eds., pp. 371-408, Ann Arbor Science, Ann Arbor, MI (1983).

99. Doraiswamy, L.K. and Kulkarni, B.D., Special Editors, "Transport Processes in Fluidized Bed Reactors," Wiley Eastern, Bombay, India (1987).

100. Yoshida, F., "Bubble Column Research in Japan," Chem. Eng. Technol., **11**, 205-212 (1988).

101. Fox, R.O. and Fan, L.T., "Stochastic Modeling of Chemical Engineering Systems--Application of the Generalized Master Equation to the Bubble Population in a Bubbling Fluidized Bed," Chem. Eng. Sci., **42**, 1345-1358 (1987).

102. Valentas, K.J. and Amundson, N.R., "Breakage and Coalescence in Dispersed Phase Systems," I&EC Fundam., **5**, 533-542 (1966).

103. Ramakrishna, D. and Shah, B.H., "A Population Balance Model for Mass-transfer in Lean Liquid-liquid Dispersions," Chem. Eng. Sci., **28**, 389-399 (1973).

104. Bajpai, R.K., Ramakrishna, D. and Prokop, A., "A Coalescence Redispersion Model for Drop-size Distributions in an Agitated Vessel," Chem. Eng. Sci., **31**, 913-920 (1976).

105. Wissler, E.H., Isbin, H.S., and Amundson, N.R., "Oscillatory Behavior of a Two-phase Natural-circulation Loop," AIChE Journal, **2**, 157-162 (1956).

106. Hsu, Y.Y. and Graham, R.W., "Transport Processes in Boiling and Two-phase Systems," pp. 261-289, Hemisphere, Washington (1976).

107. Yadigaroglu, G. and Chan, K.C., "Analysis of Flow Instabilities," Proceedings of Japan-U.S. Seminar Two Phase Flow Dynamics, Inter-University Seminar House, Kansai, Japan (1979).

108. Whitehead, A.B., Dent, D.C. and McAdam, C.H., "Fluidization Studies in Large Gas-solid Systems. Part V: Long and Short Term Pressure Instabilities," Powder Technology, **18**, 231-237 (1977).

109. Swinehart, F.M., "A Statistical Study of Local Wall Pressure Fluctuations in Gas Fluidized Column," Ph.D. Dissertation, The University of Michigan (1966).

110. Lirag, R.C. and Littman, H., "Statistical Study of the Pressure Fluctuations in a Fluidized Bed," AIChE Symposium Series, No. 116, **67**, 11-22 (1971).

111. Fan, L.T., Ho, T.C., Hiraoka, S., and Walawender, W.P., "Pressure Fluctuations in a Fluidized Bed," AIChE J., **27**, 388-396 (1981).

112. Shah, B.H., Ramakrishna, D., and Borwanker, J.D., "Simulation of Bubble Populations in a Gas Fluidized Bed," Chem. Eng. Sci., **32**, 1419-1425 (1977).

113. Ligon, J.R. and Amundson, N.R., "Modelling of Fluidized Bed Reactors-VI(a): An Isothermal Bed with Stochastic Bubbles," Chem. Eng. Sci., **36**, 653-660 (1981).

114. Ligon, J.R. and Amundson, N.R., "Modelling of Fluidized Bed Reactors-VI(b): The Nonisothermal Bed with Stochastic Bubbles," Chem. Eng. Sci., **36**, 661-671 (1981).

115. Yutani, N., Fan, L.T., and Too, J.R., "Behavior of Particles in a Liquid-solids Fluidized Bed," AIChE J., **29**, 101-106 (1983).

116. Neogi, D., Fan, L.T., Yutani, N., Nassar, R., and Walawender, W.P., "Effect of Superficial Velocity on Pressure Fluctuations in a Gas-Solid Fluidized Bed: A Stochastic Analysis," Applied Stochastic Models and Data Analysis, **4**, 13-34 (1988).

117. Mandelbrot, B.B. and Wallis, J.R., "Robustness of the Rescaled Range R/S in the Measurement of Noncyclic Long Run Statistical Dependence," Water Resources

AUTHOR INDEX

SUBJECT INDEX

Editorial Policy

This series aims to report new developments in chemical research and teaching - quickly, informally and at a high level. The type of material considered for publication includes:

1. Preliminary drafts of original papers and monographs
2. Lectures on a new field, or presenting a new angle on a classical field
3. Seminar work-outs
4. Reports of meetings, provided they are
   a) of exceptional interest and
   b) devoted to a single topic.

Texts which are out of print but still in demand may also be considered if they fall within these categories.

The timeliness of a manuscript is more important than its form, which may be unfinished or tentative. Thus, in some instances, proofs may be merely outlined and results presented which have been or will later be published elsewhere. If possible, a subject index should be included. Publication of Lecture Notes is intended as a service to the international chemical community, in that a commercial publisher, Springer-Verlag, can offer a wider distribution to documents which would otherwise have a restricted readership. Once published and copyrighted, they can be documented in the scientific literature.

Manuscripts

Manuscripts should comprise not less than 100 and preferably not more than 500 pages. They are reproduced by a photographic process and therefore must be typed with extreme care. Symbols not on the typewriter should be inserted by hand in indelible black ink. Corrections to the typescript should be made by pasting the amended text over the old one, or by obliterating errors with white correcting fluid. Authors receive 50 free copies and are free to use the material in other publications. The typescript is reduced slightly in size during reproduction; best results will not be obtained unless the text on any one page is kept within the overall limit of 18 x 26.5 cm (7 x $10^1/_2$ inches). The publishers will be pleased to supply on request special stationary with the typing area outlined.

Manuscripts should be sent to one of the editors or directly to Springer-Verlag, Heidelberg.

Lecture Notes in Chemistry

For information about Vols. 1–18 please contact your bookseller or Springer-Verlag

GPSR Compliance

The European Union's (EU) General Product Safety Regulation (GPSR) is a set of rules that requires consumer products to be safe and our obligations to ensure this.

If you have any concerns about our products, you can contact us on ProductSafety@springernature.com

In case Publisher is established outside the EU, the EU authorized representative is:

Springer Nature Customer Service Center GmbH
Europaplatz 3
69115 Heidelberg, Germany

Batch number: 10584556

Printed by Printforce, the Netherlands